MODERN METALS IN CULTURAL HERITAGE

MODERN METALS IN CULTURAL HERITAGE

UNDERSTANDING AND CHARACTERIZATION

VIRGINIA COSTA

GETTY CONSERVATION INSTITUTE, LOS ANGELES

The Getty Conservation Institute
Timothy P. Whalen, *John E. and Louise Bryson Director*
Jeanne Marie Teutonico, *Associate Director, Programs*

The Getty Conservation Institute (GCI) works internationally to advance conservation practice in the visual arts—broadly interpreted to include objects, collections, architecture, and sites. The Institute serves the conservation community through scientific research, education and training, field projects, and the dissemination of information. In all its endeavors, the GCI creates and delivers knowledge that contributes to the conservation of the world's cultural heritage.

Published by the Getty Conservation Institute, Los Angeles
Getty Publications
1200 Getty Center Drive, Suite 500
Los Angeles, CA 90049-1682
www.getty.edu/publications

Ruth Evans Lane, *Project Editor*
Sylvia Tidwell, *Manuscript Editor*
Catherine Lorenz, *Designer*
Michelle Woo Deemer, *Production*
Nina Damavandi, *Image and Rights Acquisition*

Distributed in the United States and Canada by the University of Chicago Press

Distributed outside the United States and Canada by Yale University Press, London

Printed in China

Library of Congress Cataloging-in-Publication Data

Names: Costa, Virginia, author.
Title: Modern metals in cultural heritage : understanding and characterization / Virginia Costa.
Description: Los Angeles : Getty Conservation Institute, [2019] | Includes
bibliographical references.
Identifiers: LCCN 2018045406 | ISBN 9781606066058 (pbk.)
Subjects: LCSH: Metal-work—Conservation and restoration. | Alloys. | Corrosion and anti-corrosives. | Coatings.
Classification: LCC NK6404.5 .C67 2019 | DDC 669—dc23 LC record available at https://lccn.loc.gov/2018045406

Front cover: Weathering steel (more commonly referred to as COR-TEN steel). Photo: www.pixabay.com

Illustration Credits

Every effort has been made to contact the owners and photographers of objects reproduced here whose names do not appear in the captions. Anyone having further information concerning copyright holders is asked to contact Getty Publications so this information can be included in future printings.

figs. 1.2, 1.19, 2.2, 2.7: © 2007 Michael Ashby, Hugh Shercliff and David Cebon, with permission of Elsevier
figs. 1.4, 1.6, 1.8, 1.14: © 2015 John Wiley and Sons, Inc., with permission of John Wiley and Sons, Inc. and conveyed through Copyright Clearance Center
figs. 1.15, 1.16: © 2001 ASM International, with permission of ASM International
fig. 2.9: © 1986 McGraw-Hill, with permission of McGraw-Hill Education
fig. 2.12: © 2000 ASM International, with permission of ASM International
figs. 2.13, 2.14: © 1973 Plenum Press, New York, with permission of Springer Nature
fig. 2.26: Reproduced with permission from ASTM International 2013. ASTM G46 Standard Guide for Examination and Evaluation of Pitting Corrosion, © ASTM International, 100 Barr Harbor Drive, West Conshohocken, PA 19428
fig. 4.1: © 1990 ASM International, with permission of ASM International
fig. 4.2: © 2013 Michael F. Ashby and David R.H. Jones, published by Elsevier Ltd., with permission of Elsevier
fig. 4.4: © 1967 American Society for Metals, with permission of American Society for Metals
figs. 4.5, 4.6b: © Institut national du patrimoine / Michaela Florescu
fig. 5.7: © 2017, with permission from Elsevier
fig. 6.4: © Martin Puryear, Courtesy Matthew Marks Gallery / Photo: © J. Paul Getty Trust
fig. 8.2: © 1994 Elsevier Science Ltd., Published by Pergamon / Elsevier Science Ltd., with permission of Elsevier
fig. 10.5a: Hélio Nobre / José Rosael
fig. 10.7: © 1963 Academic Press / Elsevier, with permission of Elsevier
fig. 10.9: Digital Image © The Museum of Modern Art / Licensed by SCALA / Art Resource, NY / © Robert Watts Estate, NY, 1964/2019
fig. 10.11a: Acervo História da Discoteca Oneyda Alvareugn / Centro Cultural São Paulo / SMC /PMSP
fig. 10.15: © David Mach. All rights reserved
fig. 10.16a: © 2000, with permission from Elsevier
fig. 10.16b: © Photo: Moderna Museet / Stockholm / © 2019 Judd Foundation / Artists Rights Society (ARS), New York
fig. 10.18: © 2000, with permission from Elsevier
figs. 10.22a, 10.22b: Image copyright © The Metropolitan Museum of Art. Image source: Art Resource, NY
fig. 10.22c: © Victoria and Albert Museum, London
fig. 10.23a: © photo j.m. Angles
fig. 11.4: © 2001 IPEM, with permission of Elsevier
fig. 11.7: Reprinted from *Springer Nature: Journal of Coatings Technology and Research* by N. Rani, A. K. Singh, S. Alam, et al. © 2012

CONTENTS

FOREWORD

Many of the most exciting and iconic works of modern and contemporary artists and architects—including Claes Oldenburg, Donald Judd, and Anish Kapoor; and Frank Gehry, Eero Saarinen, and Zaha Hadid—were created using metals. So, too, are household items, furniture, musical instruments, and a whole range of design objects found in museum collections.

The types of metals used, however, are very different to those used in antiquity. Developed largely for industrial purposes, modern metals comprise a vast range of new metals, alloys, and complex mixtures, often modified by a variety of surface treatments and finishes that have been adopted by architects, designers, and sculptors for their characteristically modern aesthetic and versatility.

Although modern and adaptable, many of these twentieth-century objects and structures have begun to deteriorate. Their preservation presents new and unique challenges for conservators, curators, and collectors, and knowledge about these materials is critical in creating a plan for their care and conservation. Unfortunately, there is little information available for conservation professionals on the characteristics, behavior, or technologies involved in the production of modern metals. In *Modern Metals in Cultural Heritage: Understanding and Characterization,* author Virginia Costa, a conservation scientist and expert in metals conservation, aims to fill that gap by providing conservators of cultural heritage with a deeper understanding of these materials, which will serve as the foundation for their future care.

Over the last decade, the Getty Conservation Institute (GCI), through targeted research and training initiatives, has promoted the conservation and preventive care of many modern materials—from paint, plastic, and digital prints to concrete and metals. This book was conceived following many conversations about metals between Virginia Costa and GCI conservators while she was a scholar at the J. Paul Getty Museum's Decorative Arts and Sculpture Conservation Department in 2012.

We are grateful to Virginia for generously sharing her research, her dedication to the field, and her steadfast commitment to the publication of this book. We anticipate it will be a welcome contribution to the literature available for conservators and all professionals charged with the care of modern metals in cultural heritage.

Timothy P. Whalen
John E. and Louise Bryson Director
Getty Conservation Institute

ACKNOWLEDGMENTS

To my children, Eric and Christian, with love

The idea for this book came from numerous conversations with Julie Wolfe and Sean Charette during my year as a Getty Museum scholar in 2012. The writing began more than a year later and coincided with two major events in my life: the death of my mother, Lia, and the birth of my granddaughter. The book was written largely between the daily walks to the playground and exploration of the neighborhood that Laura and I did over almost four years. Time it took for her to become a lovely girl. Time it took for the book to appear.

And during this time there were many friends and colleagues in various parts of the world, who heard me talking about the book, alternating between enthusiasm for a dream coming true and a certain weariness due to the length of the task. I am extremely grateful to all these people who contributed by sharing their experiences, providing some of the photos and information presented in the book.

In particular I thank Carla Martini for the careful reading of the manuscript and technical expertise accompanied by very pertinent suggestions. I thank Laura Moura, who, always smiling, made and remade the drawings, until they were as my mind had imagined them. I thank Anna Duer, who, like a magic trick, could get the bibliographic references I needed and Sylvia Tidwell for the very attentive editing. I thank my husband, Philippe Allongue, who read each chapter trying to improve the form, and who, with his critical comments, gave me (after a brief moment of despair, I recognize) the necessary strength that took me to position myself even more firmly on what I wanted.

Finally, I thank the many people at the Getty who helped bring this book to the light: Jeanne Marie Teutonico, Cynthia Godlewski, and Alison Reilly at the Getty Conservation Institute, and Ruth Evans Lane, Catherine Lorenz, Michelle Woo Deemer, and Nina Damavandi in Publications.

INTRODUCTION

> But if exact science is used to illuminate empirical experience, and if experience is used to temper the extrapolation of the simple ideal systems of the scientist, then indeed we have real knowledge.
>
> —Cyril Stanley Smith, *A Search for Structure: Selected Essays on Science, Art and History*

Since the nineteenth century, the proliferation of new metallic alloys, developed for specific industrial and commercial purposes, has greatly expanded the class of materials commonly referred to as "modern" or "contemporary" metals by conservators. New fabrication and working processes, the use of new materials in alloying, and precise formulations for specific functions have produced a broad category of complex materials that are usually described by their main constituent elements and numerical grades rather than by the comparatively straightforward nomenclature that was used to describe the bulk of ancient or historical metals in museum collections.

Modern metals—such as aluminum alloys, stainless steels, and metallic coatings—are used in a variety of contexts including decorative art, architecture, industrial manufacture, scientific instrumentation, and everyday life and thereby constitute a considerable part of humanity's cultural heritage. They have not yet been systematically studied, characterized, or described in the conservation literature and are often simply designated "metal" on museum labels placed next to artifacts. Moreover, the few and quite diverse conservation-related resources in the literature very often lack appropriate materials characterization, reasons for choosing a particular intervention, a systematic approach, and transparency in the presentation of results. Because modern metals have become more prevalent in collections and require specific conservation attention, the shortage of professionals capable of understanding and executing suitable conservation protocols for them is becoming a pressing issue.

An important goal of conservation science is to disclose and make accessible the information that an artifact carries. In order to conserve an artifact, it is of fundamental importance to know and understand both its immaterial and material aspects—that is, to characterize it fully. A complete knowledge of materials and techniques used for the elaboration of the object is also fundamental for assessing its conservation condition and making appropriate decisions about intervention. The characterization is accomplished primarily through a detailed visual inspection, followed by an appropriate investigation for confirmation and documentation. The knowledge thus acquired is crucial for validating maintenance procedures, especially in long-term conservation.

Artifacts made of modern metals are changing the field of metals conservation. The subtle chemical compositions and structural nuances of these new materials must be disclosed and recorded, but there are no established strategies for the conservation of this heritage. Although related technical literature is vast and detailed, only professionals with sufficient background can make use of it to establish protocols in particular cases. Moreover, modern metals are composed of so many small and precise amounts of alloying elements that most nondestructive analytical techniques are unable to distinguish the tiny metallurgical differences that exist within a given class of alloys, and conservation professionals can hardly describe them in a comprehensive way.

This book aims at providing basic knowledge to professionals dealing with modern metals—conservators, conservation scientists, curators, artists—to alert them to the specificity of these materials in relation to their conservation. By presenting a comprehensive overview of a selection of modern alloys and coatings, the present volume aims to build upon the information available in technical sources to shed more light on the characteristics, production technologies, applications, and conservation of these materials. A glossary at the end of the book provides definitions of key terms. This work intends to bridge the gap between published technical information and the people in charge of metallic heritage, serving as a cornerstone reference for the field as well as a valuable resource for training and education programs. This book does not, however, recommend, describe, or even reference treatment method.

The volume is divided into three parts. Part 1 deals with basic concepts about metallic alloys: structure, properties, and characterization. These topics are central to addressing the conservation of modern alloys, whose nuances of composition and structure are determinant for their conservation. Chapter 1 introduces the different metallic structures and describes their evolution when subjected to various metallurgical processing steps, such as casting, forming, alloying, and heat treating. The metallic microstructure of an artifact, together with the alloy composition, constitute historical testimony, because the microstructure results from the successive elaboration steps that were employed in the object's manufacture as well as from its use. Characteristic properties —mechanical and chemical—of metallic alloys are presented in chapter 2. This knowledge is key to characterizing metallic-alloy artifacts as well as assessing their conservation state. Chapter 3 deals with the characterization of metallic-alloy artifacts, emphasizing the importance of a preliminary visual inspection and the necessity of obtaining information about the artifact's history in order to raise precise questions and formulate an effective analytical approach.

Part 2 covers the most important modern alloys. They include aluminum alloys, whose huge variety covers applications as simple as beverage cans and as complex as aircraft parts; copper alloys, whose wealth of types goes beyond traditional versions of brass and bronze; stainless steels, whose particular surface properties must be better understood; titanium alloys, currently often employed in architecture and conservation; weathering steels, with their highly nuanced properties that must be considered in design projects; and zinc alloys, which are of general interest for die casting. In each chapter, after a brief historical introduction, the metallurgical background of each alloy type is explained. Next, for the most frequently used commercial types, the designation, properties, and applications are addressed, with links to standards. There is also a section on deterioration mechanisms of each of these alloys, followed by a discussion on characterization. A final section provides a comprehensive selection of important conservation publications related to the alloy.

Part 3 deals with inorganic coatings, which are divided into two groups: metallic coatings (chap. 10) and conversion coatings (chap. 11). Metallic coatings include those produced by electroplating, electroless processes, and hot dipping, while those obtained by conversion comprise anodizing, chromate, and phosphate coatings. In every case, the principles and processes are presented, followed by descriptions of the structure, properties, and applications of the most commonly used coatings. As in the chapters on the alloys, a final section lists publications related to the coatings in the conservation field.

PART I | METALS AND ALLOYS
NATURE, PROPERTIES, AND CHARACTERIZATION

CHAPTER 1 | THE NATURE OF METALS

The range of metallic materials covers most of the periodic table of the elements (fig. 1.1). Rather than defined, they are recognized as a group by the properties they have in common. ASM International, formerly known as the American Society for Metals (ASM), describes a metal as "an opaque lustrous elemental chemical substance that is a good conductor of heat and electricity and, when polished, a good reflector of light." Most elemental metals are malleable and ductile and are, in general, denser than the other elemental substances (Reardon 2011, 446).

To know the features that various metals have in common that can determine these properties, we have to look deep inside the metal itself and examine its structure—that is, the arrangement of its internal components from atomic to microscopic scale, which is the level at which properties are defined. Figure 1.2 represents microstructural features in metals on different length scales and the properties that they determine.

Such features are common to all of the different metallic alloys and can be seen as constituting a generic matrix. Nearly all alloys are simply modifications of the basic recipe. Knowledge of this matrix enables understanding of how the structure of alloys evolves during processing and thereby enables alteration of the material properties.

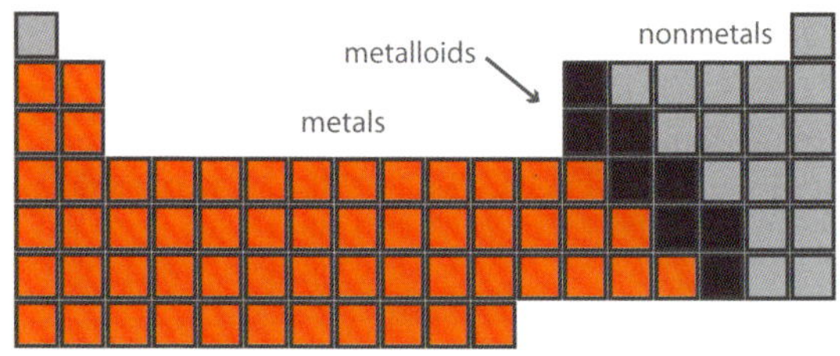

Figure 1.1 | The periodic table of elements with the metals group highlighted (orange).

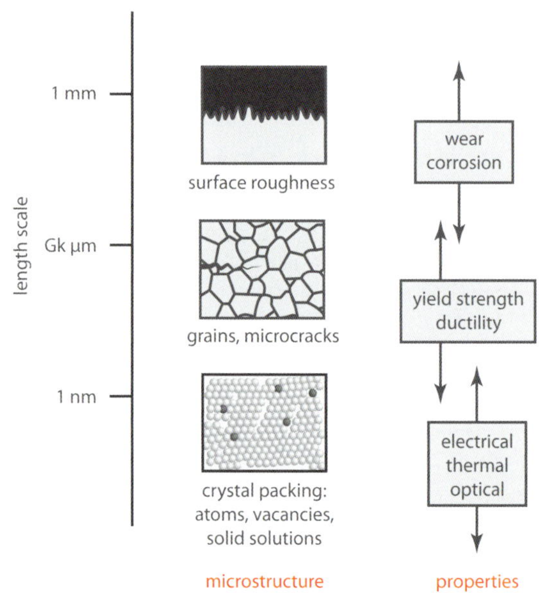

Figure 1.2 | Microstructural features in metals at different length scales and the properties that they determine (adapted from Ashby, Shercliff, and Cebon 2007, fig. 19.1).

STRUCTURE

Metallic Bonding

Many of the properties of materials can be predicted from the interatomic forces that bind the atoms together (Campbell 2008). The bonding occurs due to the tendency of the atoms to reach a stable electronic structure, like the one of inert gases, by completely filling the outermost electron shell. Metallic materials have only a few valence electrons, which are not bound to any particular atom in the solid, drifting more or less freely throughout the entire metal. In metallic materials, the cohesion between atoms comes from electrostatic forces because the metal atoms give their electrons (from the outer shell) to the surrounding environment, forming an ensemble of positive charges within a sea of electrons that are negatively charged (fig. 1.3). In other words, metal atoms do not form rigid chemical bonds, and the electrons spread freely inside the matter, a property that gives metals their good electrical and thermal conductivity. Furthermore, the nonrigid characteristic of the metal bonds explains the mechanical properties: thanks to the slip of atomic planes, when under load, most metals and their alloys exhibit significant degrees of permanent deformation before fracture occurs.

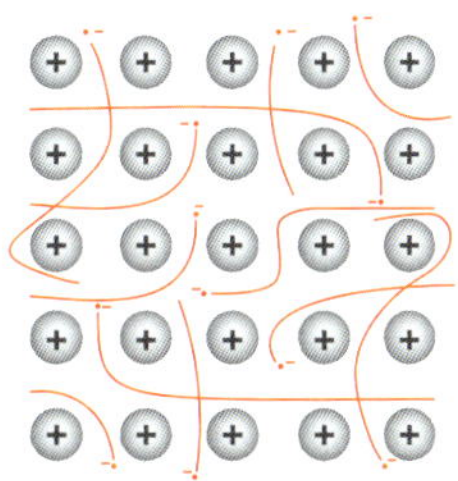

Figure 1.3 | Metallic bonding.

Crystalline Structure

On an atomic scale, metals are crystalline: their constituent atoms are stacked together in a repeating pattern, forming a three-dimensional periodic network (Campbell 2008). In fact, atoms tend to behave like miniature ball bearings and to pack together as tightly as possible to minimize free space. Nearly all metals crystallize within the three main structural categories: body-centered cubic (BCC), face-centered cubic (FCC), and hexagonal close-packed (HCP) (fig. 1.4).

Although metals can crystallize within more complex forms, the three close-packed structures described above are very important because they favor the slipping of atomic layers. When a metal undergoes mechanical deformation, its crystalline structure rearranges to accommodate the deformation. Such rearrangement takes place by slip, or sliding, on the close-packed planes and leads to plastic (permanent) deformation (fig. 1.5). Therefore, the greater the number of such planes available in a given crystalline system, the greater the capacity of plastic deformation. This is a unique property of metals: they can be bent and worked into a desired shape. Metals crystallizing within the FCC structure (Cu, Au, Al) are capable of moderate to extensive plastic deformation. BCC metals (Fe, Cr, Mn) have a structure that is

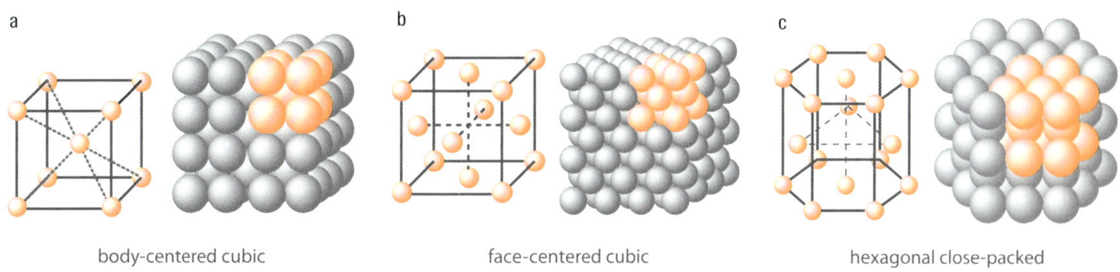

Figure 1.4 | Crystal structures of most of the common metals: Body-centered cubic (a); face-centered cubic (b); and hexagonal close-packed (c). Each is represented as a reduced-sphere unit cell (left) and as an aggregate of many atoms (right) (aggregate representation adapted from Callister 2015, figs. 3.1c, 3.2c, and 3.4b).

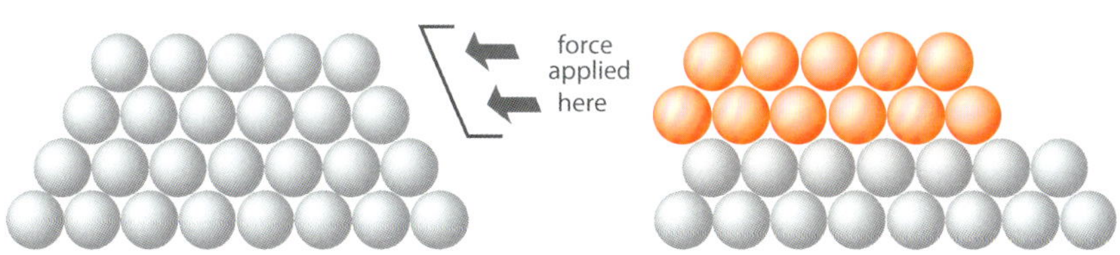

Figure 1.5 | The position of atoms in a metallic crystal before (left) and after (right) slip has occurred.

less compact, which causes them to fracture prior to reaching their maximum possible plastic deformation. In general, the plastic deformation of HCP metals (Zn, Mg, Ti) is much more limited. They have poor to only moderate room-temperature ductility, often requiring heating to elevated temperatures, at which slip becomes much easier, prior to forming operations.

In fact, the slip of atomic planes under a given applied force is made easier by the presence of imperfections, or structural defects, contained in a considerable number in the crystalline lattice. Among defects, dislocations are the most important ones. Virtually all crystalline materials contain some dislocations that were introduced during solidification, as a consequence of thermal stresses that result from rapid cooling, or by atomic mismatch due to the presence of alloy elements or phase transformation. They are characterized by regions of atomic misfit, where atoms are locally not properly surrounded by neighboring atoms, as elsewhere. As metals plastically deform, dislocations in the crystal matrix move around in response to the external applied forces, allowing the crystal to deform with a minimal energy input (fig. 1.6). In fact, practically all mechanical properties of metals are, to a significant extent, controlled by the behavior of dislocations.

Microstructure

Metals do not have a continuous crystalline structure but are composed of a collection of small crystals or grains and are therefore designated as polycrystalline (Campbell 2008). These small crystals originate during solidification, as represented in figure 1.7. As the temperature decreases from the melting point, small solidification nuclei form at various positions in the melt. These have a random crystallographic orientation, as indicated by the square grids, and they grow by the successive aggregation of atoms from the surrounding matter in the liquid phase. At some point the boundaries of adjacent grains impinge on one another as the solidification process approaches completion.

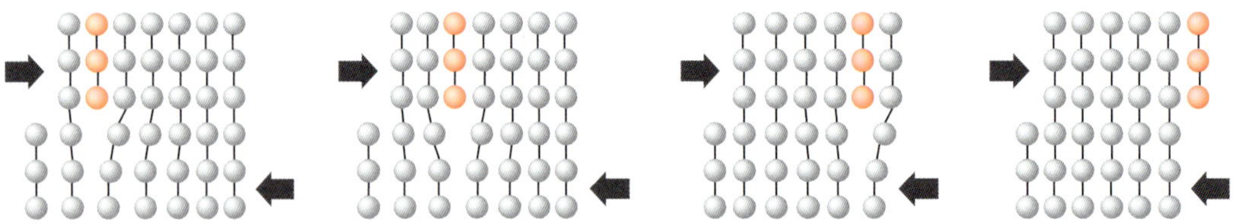

Figure 1.6 | The motion of a dislocation (adapted from Callister 2015, fig. 8.3).

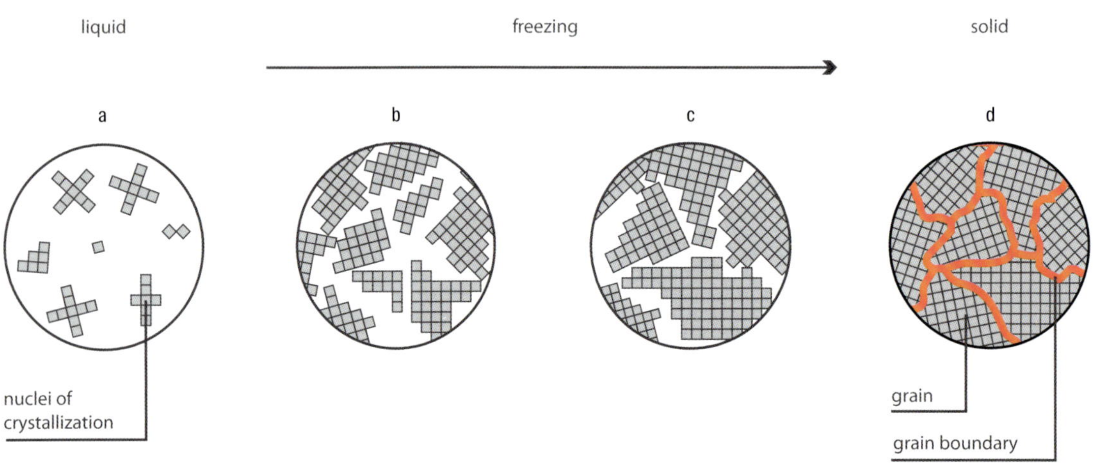

Figure 1.7 | Solidification stages of a molten metal (small squares represent unit cells with different crystallographic orientation): Nucleation of crystals at random sites in the molten metal (a); growth of crystals as solidification continues (b and c); and solidified metal with individual grains and grain boundaries where the crystallographic orientation changes direction (d).

As a result of this process, the crystallographic orientation varies from grain to grain, and this variance results in some atomic mismatch at the interface where two grains meet, which is called a *grain boundary*. Grain boundaries are quite complicated regions because they are not only extended structural defects but also the stage for the segregation of chemical impurities. Grain boundaries often play an important role in the mechanical properties and corrosion behavior of metals.

Grains and their boundaries—the microstructure—can be revealed by optical microscopy observation after careful and meticulous surface preparation (fig. 1.8). The specimen is first ground and polished to a mirrorlike finish, and the microstructure is revealed by applying an appropriate chemical treatment called etching. The grain size is typically 10–100 μm.

a

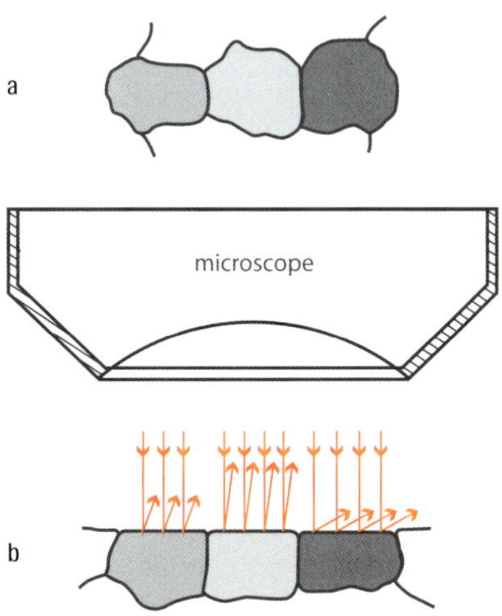

b

Figure 1.8 | Polished and etched grains as they might appear when viewed with an optical microscope (a); section of grains, showing how the etching characteristics and the resulting surface texture vary from grain to grain because of differences in crystallographic orientation (b) (adapted from Callister 2015, fig. 5.19).

EVOLUTION OF MICROSTRUCTURE

Manufacturing processes are central to achieving the target properties in a material. Shaping processes such as casting, molding, and forming always impose a thermal history and often a deformation history. These processes do more than create a shape; they also control the evolution of the internal microstructure upon which properties depend. Subsequent processing, such as polishing or welding, can further change the microstructure and properties locally. This ability to develop different property profiles simply by modifying chemistry, forming, and heat treatment gives metallic alloys a great versatility; overall, they cover a wide domain of the key structural properties.

Some of the manufacturing processes that determine the structure and properties of metals are introduced below.

Casting

In casting, a metal previously heated above its melting point is poured into a mold, where it cools. As explained above, the solidification process occurs by nucleation and growth of individual nuclei, starting usually on the walls of the mold or on foreign particles in the melt itself. Each original nucleus is the origin of a grain in the solid. Control of the final grain size is very important to obtaining the desired properties of the solid. This goal is achieved by choosing an appropriate cooling rate to control the initial number of nuclei in the melt and the rate of their growth, as well as through the addition of substances, called inoculants, that stimulate nucleation (fig. 1.9).

Metals usually contain impurities and are often mixed with others to form alloys. Although in the molten metal all of these foreign elements are uniformly distributed, their solubility in the solid phase decreases, so they are rejected ahead of a growing crystal. Since the last parts to solidify are the grain boundaries, concentration gradients may be set up inside the grains. This segregation can later be a source of problems, such as embrittlement or enhanced corrosion susceptibility at the grain boundaries. Castings are therefore often held at a high temperature for prolonged periods to favor a more uniform distribution of solutes within the solid (fig. 1.10).

Casting is used to produce objects in their final form or to produce ingots that will be further worked by other forming processes. Casting methods include die casting, sand casting, and the lost-wax process (see table 3.2).

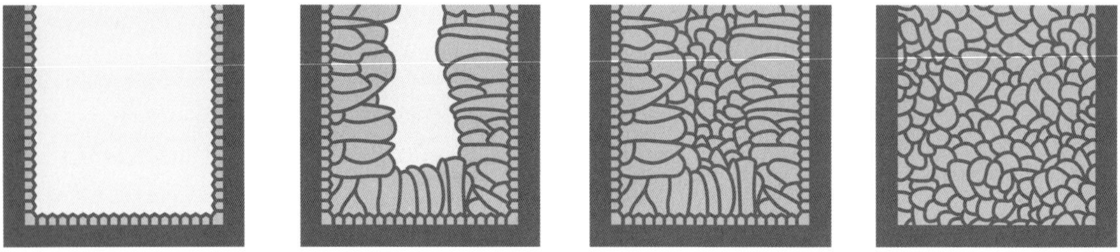

a b c d

Figure 1.9 | Structural zones forming in a casting: Chill zone—at the mold walls, metal cools rapidly, producing a solidification "skin" of fine equiaxed grains (a); columnar zone—grains grow preferentially toward the center of the mold (b); equiaxed zone—as the driving force of the heat transfer decreases away from the mold walls, the grains become equiaxed and coarse (c); and homogeneous structure obtained by using nucleating agents (d).

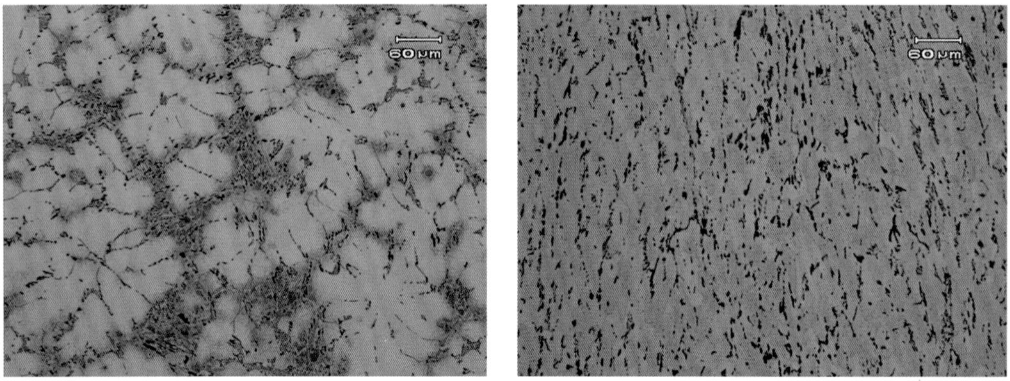

a b

Figure 1.10 | Microstructure of aluminum AA2618 alloy showing dendritic microsegregation in the cast and forged condition (a); and dissolution of interdendritic segregation after solution heat treatment at 535°C for 4 hours (b). Photo: Nicoli dos Santos Simões, LAMEF/UFRGS.

Metalworking Processes

Metalworking processes shape cast structures to produce what are called *wrought structures* (Campbell 2013a). Bulk deformation processes like rolling, forging, extrusion, and drawing involve large amounts of plastic deformation. In addition to shaping the metal, such processes are used to refine the grain size, break up segregation, and eliminate defects such as pores and cavities, which are often present in cast alloys.

Processes used for the initial breakdown of cast ingots are always conducted hot, where greater deformation is possible and lower forming energies are required. They are conducted at temperatures above the recrystallization temperature, which is approximately 0.5 T_m (melting temperature). During hot working, the hardened and distorted grain structure produced by deformation is almost simultaneously eliminated by the formation of new strain-free grains as a result of recrystallization (fig. 1.11a). Secondary processes, which are used to produce the final product shape, are conducted either hot or cold. Cold working processes are conducted at or near room temperature. During the process, the grains distort, the grain boundaries align in the direction of metal flow, and a well-defined fiber-like pattern develops (fig. 1.11b). This results in directional variation of mechanical properties between the worked direction and the other directions (the so-called anisotropic behavior of the deformed material). Cold forming produces improved surface finish and increases the strength of the product through strain hardening.

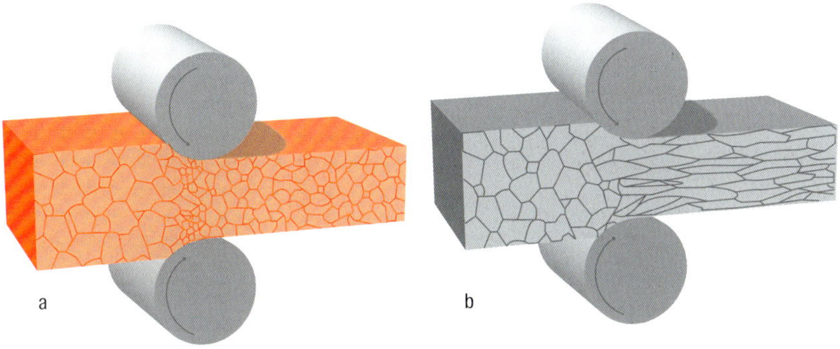

Figure 1.11 | Changes in grain structure during hot rolling (a) and cold rolling (b).

Alloying

Few metals are used in their pure chemical state. They are usually mixed with one or several other chemical elements in variable amounts to produce alloys with improved properties (Baker 2001). Alloying elements are usually added to the liquid metal—that is, when it is above the melting point—where they are dissolved easily. During solidification the transition from liquid to solid state shortens atomic distances, and the solubility of the added element (solute) inside the crystal lattice of the base metal (solvent) decreases. At room temperature, the solubility varies depending on the chemical elements that are combined. For instance, it can range from as little as 0.007% carbon in iron for a carbon steel to as much as 100%—complete solid solubility—for the copper-nickel system. The insertion of solute atoms introduces strains in the crystal lattice, creating barriers against the motion of dislocations and causing some hardening and strengthening of the alloy (solid solution strengthening) (fig. 1.12).

The final arrangement of the alloying elements in the solvent lattice determines the properties of the metallic alloy. When there is enough room for all solute atoms inside the crystal lattice, in either vacancies or interstices or both, a solid solution is formed. In this case the material is no longer a pure metal but an alloy, constituted by a single phase—that is, a region of material that has uniform chemical and physical properties (fig 1.13a). Even if such changes cannot be discerned by examination with an optical microscope, their effects can be measured by the change in properties.

When the alloy contains more solute atoms than the host metal can dissolve, the surplus will split up into more than one phase and/or chemical compound. In a few cases, the result is a solid containing two phases: one rich in the solvent element, containing just enough of the solute atoms, and the other rich in the surplus (fig. 1.13b). In most alloy systems, however, the surplus atoms of the alloying element separate out as chemical compounds, such as CuA_{12} in Al-Cu alloys or Fe_3C in carbon steels. Since such compounds are hard and

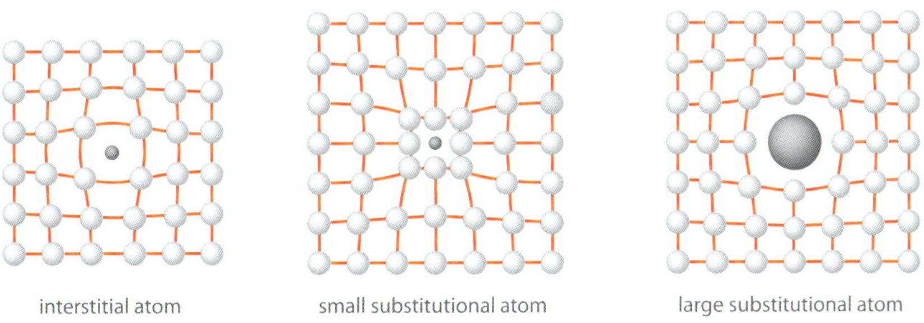

Figure 1.12 | Lattice distortions caused by solute addition.

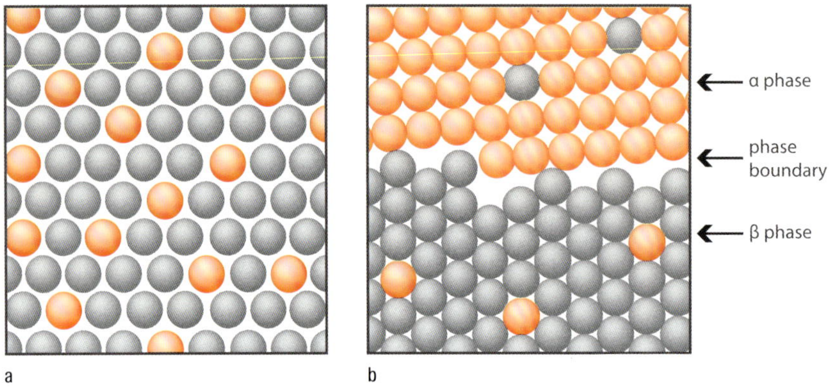

Figure 1.13 | Atomic arrangement in an alloy. One phase (a): Solid solution of B (orange atoms) in A (gray atoms); two phases (b): solid solution of B in A plus solid solution of A in B.

not easily cut by dislocations, they can yield large increases in strength when they are finely dispersed throughout the alloy, and they can weaken locally when they are segregated into specific areas.

All these features can be predicted from the phase diagrams, which are graphic maps of the equilibrium phases associated with various combinations of the state variables—that is, temperature and composition. Phase diagrams illustrate the evolution of the different phases and associated microstructure that results from variation in temperature and composition. Even though such diagrams are made considering equilibrium (ideal) conditions, they provide a baseline for inferring the actual structure of the alloy.

There are phase diagrams for almost all alloy systems. An alloy system consists of all the alloys that can be made with a given set of components. For example, the Cu-Ni system comprises all the alloys that can be made from copper and nickel. For binary alloys—those with two components—the composition of the alloy is plotted on the x-axis (in weight or atom %), and the temperature is plotted on the y-axis. The diagram is simply a two-dimensional map (drawn up from experimental data on the alloy system) that shows where the various phases are in the composition-temperature field. The field where the alloy is liquid is designated by *L*, and the solid phases are often designated by Greek symbols, such as alpha and beta.

Figures 1.14–1.16 illustrate the usefulness of phase diagrams to predict the structure (and consequently the properties) resulting from the addition of different elements to copper under equilibrium conditions of slow cooling. The copper-nickel diagram is presented in figure 1.14. Both metals are completely miscible, and whatever the composition of the alloy, nickel atoms substitute directly into the lattice in place of copper atoms. This substitutional phenomenon results in a single-phase solid solution (alpha) of nickel in copper. The crystalline structure, though stronger than pure copper because of lattice strain, is still FCC. Consequently, the slip characteristics remain very good. Single-phase solid solution alloys of copper, therefore, retain good deformation characteristics despite very significant increases in strength.

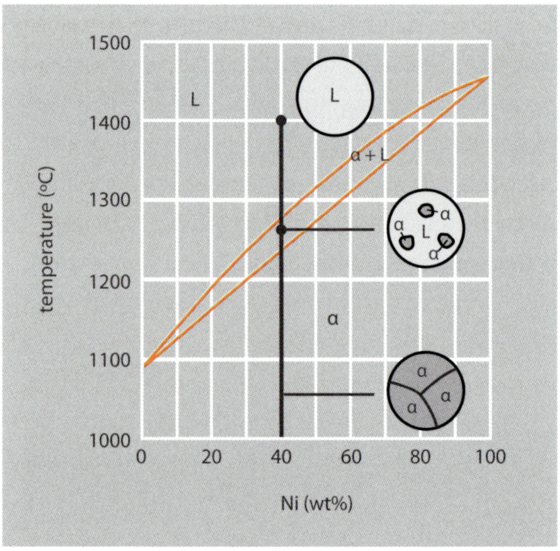

Figure 1.14 | Partial phase diagram for the copper-nickel alloy system showing the structural changes during solidification at a 40% Ni alloy (vertical gray line). The top circle depicts the liquid phase; the middle circle shows the nucleation of crystals; and the bottom circle shows the granular structure of the solidified alloy (adapted from Callister 2015, fig. 10.3a).

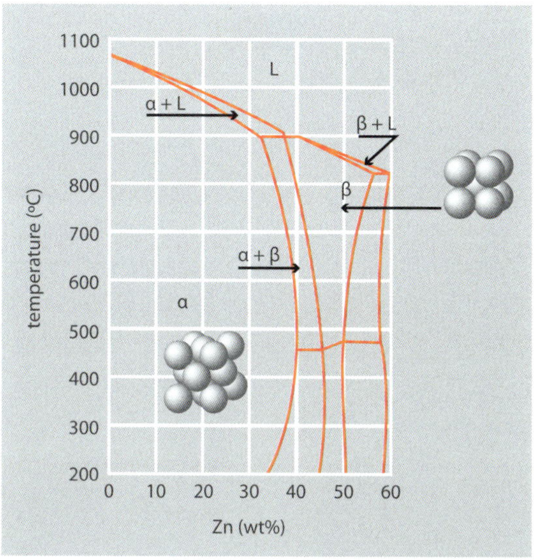

Figure 1.15 | Partial copper-zinc phase diagram. Depending on the zinc content, the alloy consists of one or more phases (adapted from Davis 2001, fig. 10).

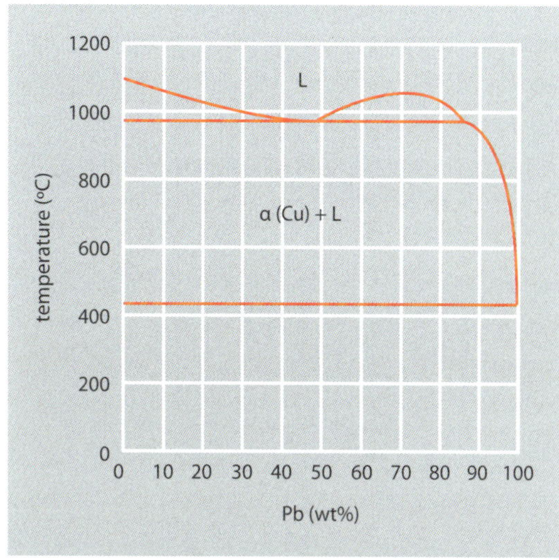

Figure 1.16 | Copper-lead phase diagram. There is no solubility of lead in solid copper. Lead remains molten in solid copper above 326°C (adapted from Davis 2001, fig. 8a).

A similar behavior can be partially observed in the case of copper-zinc alloys (fig. 1.15). When up to about 35% zinc is added to copper, it dissolves to form a solid solution of uniform composition designated alpha (FCC). Further increase in zinc content produces a mixed structure, containing an additional solid solution of higher zinc content, known as the beta phase (BCC). Brasses containing between about 35%–45% zinc consist of mixtures of these two phases and are known as alpha-beta or duplex brasses. These materials are known as polyphase alloys and are characterized by higher strength, hardness, and wear resistance than alpha alloys. An additional characteristic of the polyphase alloys, compared to the single-phase materials, is that their properties can vary with temperature. Such alloys are often heat treatable.

Finally, the copper-lead phase diagram is depicted in figure 1.16. It can be seen that lead is nearly completely rejected by the copper lattice at low temperatures. Nonetheless, with special metallurgical techniques, lead may be captured between the crystals of copper-based material during solidification, so that it remains dispersed in the matrix.

To conclude, whenever one needs to know or estimate the properties of a metallic alloy, the first order of business is to reach for the phase diagram. It will provide the baseline from which other (nonequilibrium) structures can be inferred.

Heat Treatment

Because the solubility of the elements in the alloy is dependent on temperature, it is possible to control the mode of occurrence of the alloying elements in the structure of metallic alloys by performing heating and cooling treatments under given conditions. For every alloy, specific combinations of time and temperature are required in order to obtain or improve particular properties such as workability, strength, relief of residual stresses, dimensional stability, and stabilization of properties so that they do not change with time (Callister 2015; Campbell 2013d).

Heat treatment is a process performed in at least two steps. The first part consists of heating the alloy and maintaining it for a given period of time at a temperature that is usually just high enough to solubilize the alloying atoms in the solvent matrix. The second part involves cooling at a rate chosen to obtain a given structure. Slow cooling allows constituents to accommodate themselves during solid state transformations, while rapid cooling will prevent some reactions from occurring. Therefore, depending on the cooling rate, the structure (and properties) may be completely different for the same alloy.

Although all metallic alloys respond to some type of heat treatment, the response is by no means equivalent. While almost any alloy can be softened by means of a suitable heating-and-cooling cycle, the number of alloys that can be strengthened or hardened by heat treatment is far more restricted. Practically all of the ferrous alloys respond to at least one type of heat treatment, undergoing phase transformations that allow for a very wide range of properties. This is the principal reason why these alloys have been so extensively used in the manufacturing sector during the twentieth century. Many nonferrous alloys and some stainless steels do not undergo phase transformations like those in steels, and they can be strengthened to various degrees by a process called precipitation hardening (PH). A short description of the most widely used heat-treatment processes follows.

Annealing

Annealing is primarily used for the restoration of cold-worked or heat-treated alloys to their original properties. Basically, it consists of heating to and holding at a specific temperature, followed by cooling at a low rate, usually in air or a turned-off furnace. For nonferrous alloys, it is a common treatment for the softening of metals previously hardened by cold work,

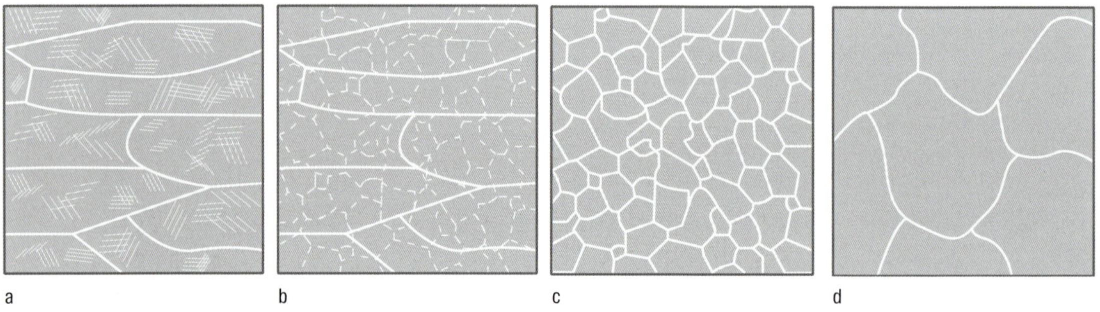

Figure 1.17 | Grain evolution through annealing: Initial cold-worked grains, showing slip lines (a); nucleation of new grains starts during annealing (b); and subsequent stages in the development and growth of new grains during annealing (c and d).

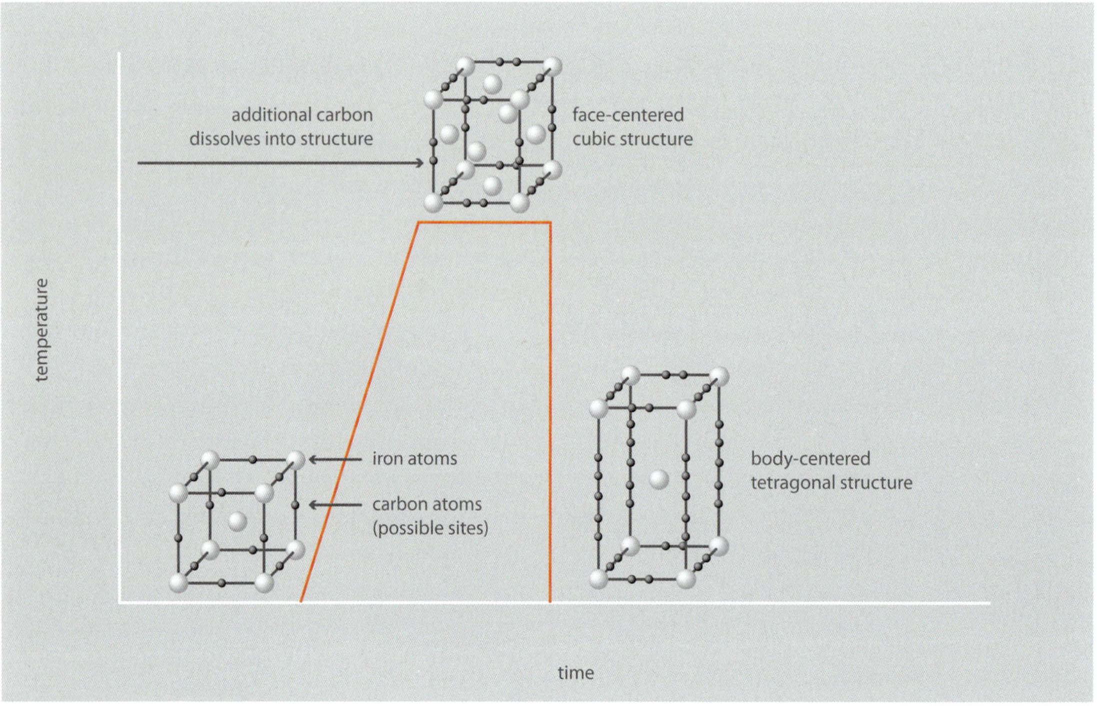

Figure 1.18 | Temperature-versus-time plot showing lattice transformations during quenching of ferrous alloys: The BCC lattice turns into FCC by heating, and the latter turns into BCT by rapid cooling. The final structure is martensite.

which have lost their malleability, and for controlling the grain structure in the final product (fig. 1.17). Annealing is also used for the homogenization of as-cast structures.

In ferrous alloys, annealing time-temperature cycles vary widely in maximum temperature and cooling rate. Full annealing denotes heating to high temperatures to promote phase transformation; stress-relief annealing is performed at lower temperatures, when the only purpose is to release stresses.

Quenching

Quenching is a heat-treatment process used mainly to strengthen iron-carbon alloys. It consists basically of two steps: heating the alloy to a temperature high enough to solubilize the carbon atoms in the austenite and then rapidly cooling it, a procedure that results in an almost instantaneous structural transformation—from FCC to body-centered tetragonal (BCT)—that prevents carbon diffusion (fig 1.18). An alloy with this structure, called martensite, is extremely hard and brittle. Typically, quenching media are liquids or gases: oil; water; water with salt or caustic additives; and inert gases including He, Ar, and N.

Most steels cannot be used in the as-quenched condition, because they are very hard and brittle and contain internal stresses. In order to solve this problem, steels are submitted to a subsequent treatment called tempering. The previously hardened material is usually heated to a temperature below the lower critical temperature and is cooled at a suitable rate. Steels are tempered by reheating after hardening to obtain specific values of mechanical properties, relieve quenching stresses, and ensure dimensional stability.

Precipitation Hardening

Many nonferrous alloys and some stainless steels undergo precipitation hardening to enhance strength. Even though the treatment uses procedures similar to those for hardening steel (quenching and tempering, discussed above), the mechanisms by which precipitation hardening hardens and strengthens are entirely different.

In precipitation hardening, the alloy is heated to a temperature high enough to take a significant amount of an alloying element into solid solution (fig. 1.19a). It is then rapidly cooled to room temperature, a step that traps the alloying elements in solution (fig. 1.19b). On reheating to an intermediate temperature, the host metal rejects the alloying element in the form of fine precipitates that create strains in the lattice (fig. 1.19c). These precipitates act as barriers to the

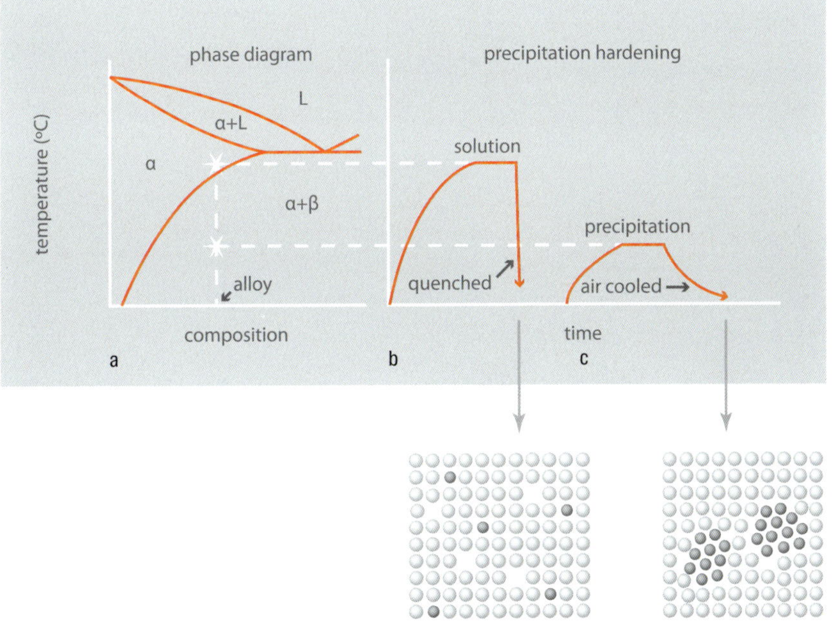

Figure 1.19 | Phase diagram showing decreasing solubility with temperature (a); thermal profile for heat treatments that produce precipitation hardening, illustrating microstructural changes: solution (b) and precipitation (c) (adapted from Ashby, Shercliff, and Cebon 2007, fig. 19.7).

motion of dislocations and provide resistance to slip, thereby increasing the strength and hardness.

Because the precipitation process strongly depends on time and temperature, it is also called aging, and the property improvement is known as age hardening. Some alloys, especially aluminum alloys, harden and become stronger over a period of time at room temperature; this process is therefore called natural aging. If the treatment is carried out above room temperature, it is called artificial aging.

In conclusion, the microstructure of alloys reflects the origin and subsequent history of metallic products since it results from the various mechanical and thermal treatments involved in their manufacture. Differences in microstructure result in great variances in strength, toughness, and corrosion resistance among alloys of similar composition and can frequently be used to explain or predict the characteristics and performance of alloys.

REFERENCES AND BIBLIOGRAPHY

Ashby, M., H. Shercliff, and D. Cebon. 2007. *Materials Engineering, Science, Processing and Design*. Oxford, MA: Elsevier.

Baker, H. 2001. "Principles of Alloying." In *Alloying: Understanding the Basics*, edited by J. R. Davis, 3–18. Materials Park, OH: ASM International.

Callister, W. D. 2015. "Phase Transformations." In *Fundamentals of Materials Science and Engineering*, 323–64. New York: John Wiley and Sons.

Campbell, F. C., ed. 2008. *Elements of Metallurgy and Engineering Alloys*. Materials Park, OH: ASM International.

———. 2013a. "Bulk Deformation." In *Metals Fabrication: Understanding the Basics*, edited by F. C. Campbell, 103–62. Materials Park, OH: ASM International.

———. 2013b. "Casting." In *Metals Fabrication: Understanding the Basics*, edited by F. C. Campbell, 47–102. Materials Park, OH: ASM International.

———. 2013c. "Forming." In *Metals Fabrication: Understanding the Basics*, edited by F. C. Campbell, 163–212. Materials Park, OH: ASM International.

———. 2013d. "Heat Treatment." In *Metals Fabrication: Understanding the Basics*, edited by F. C. Campbell, 271–324. Materials Park, OH: ASM International.

Davis, J. R. 2001. *Copper and Copper Alloys*. ASM Specialty Handbook. Materials Park, OH: ASM International.

Leo, P., and E. Cerri. 2013. "Pure 7000 Alloys: Microstructure, Heat Treatments and Hot Working," in *Aluminium Alloys: New Trends in Fabrication and Applications,* edited by Z. Ahmad, 258–60. London: IntechOpen.

Reardon, A. C. 2011. *Metallurgy for the Non-Metallurgist*. 2nd ed. Materials Park, OH: ASM International.

CHAPTER 2 | PROPERTIES OF METALLIC ALLOYS

A property is an inherent characteristic of a material in terms of the nature of its response to a particular imposed stimulus and the magnitude of that response. For example, a specimen subjected to forces will experience deformation, or a polished metal surface will reflect light. Among the many properties of solid materials—including electrical, thermal, magnetic, and optical—two will be discussed in detail in this chapter because of their importance to metallic alloys. These are the mechanical properties, which relate deformation to an applied load or force, and the deteriorative properties, which are a function of the chemical reactivity of materials to environmental factors, including moisture and pollutants.

MECHANICAL PROPERTIES

Strengthening Metals

The strength of a piece of metal is determined by its resistance to plastic (permanent) deformation upon application of a load. Since plastic flow in crystalline solids is allowed by the motion of crystal defects, especially dislocations (see chap. 1, "Microstructure," and fig. 1.6), metal alloys can be strengthened by increasing the interference to the slip process, making it more difficult for the dislocations to move. Practically all strengthening techniques rely on this simple principle: Restricting or hindering dislocation motion renders a material harder and stronger (Reardon 2011b; Callister 2015a).

Strengthening can be improved by refining—that is, by reducing grain size. Adjacent grains, which have random crystallographic orientations, are separated by a grain boundary. During plastic deformation, slip must take place across grain boundaries, and the traveling dislocations must change their direction of motion (fig. 2.1). This process is made more difficult as the crystallographic misorientation increases, a result of reduction in grain size. A fine-grained material is harder and stronger than a material with coarse grain, since the former has a greater grain boundary area per volume unit, which efficiently impedes dislocation motion.

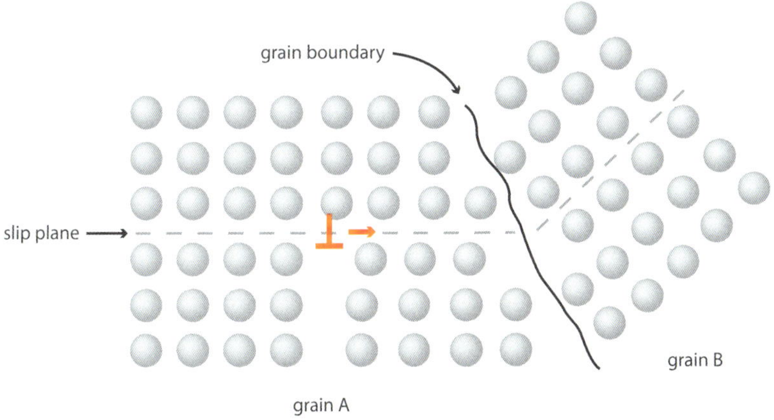

Figure 2.1 | A dislocation moving in grain A, which encounters an obstacle when approaching the grain boundary; because of lattice misalignment, the dislocation cannot move into grain B but is trapped at the grain boundary (adapted from Callister 2015a, fig. 8.14).

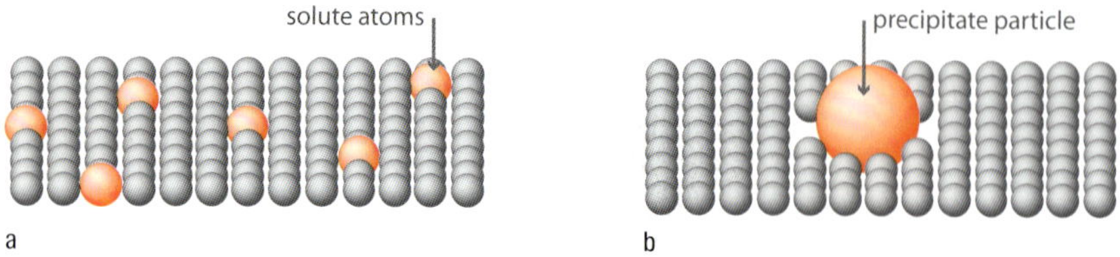

Figure 2.2 | Solid solution strengthening: View of a slip plane from the perspective of an advancing dislocation. Introduction of alloying atoms in the crystal lattice creates obstacles to dislocation motion—atom sized in the case of solid solution strengthening (a), and larger obstacles in the case of precipitation hardening (b) (adapted from Ashby, Shercliff, and Cebon 2007, fig. 6.17).

Another strengthening mechanism is work hardening, whereby the metal is plastically deformed, generating new dislocations and causing their density to increase from around 10^{10} m/m^3 to as much as 10^{17} m/m^3 or more. The result is that the motion of a dislocation is significantly hindered, and the stress required to further deform the metal is consequently increased. Sometimes this mechanism is also called cold working, because the temperature at which deformation takes place is "cold" relative to the absolute melting temperature of the metal. During manufacture, it can be problematic, because it requires the intermediate step of heating the piece to remove the accumulated dislocations before continuing. It is also useful, however: it is a very effective strengthening method, particularly for alloys that cannot be hardened by heat treatment.

Metals can be also strengthened in many ways by alloying to different degrees. The simplest method is to add a foreign element that forms a solid solution with the solvent metal. Since the added atoms never exactly match those in the original lattice (see fig. 1.12), the slip plane becomes "rough," thereby adding additional resistance to the motion of dislocations (fig. 2.2a). Moreover, when the amount of added atoms exceeds the limit of solid solubility, as the temperature is decreased, the excess alloying atoms try to unite with atoms of the base metal to form small regions of intermetallic compound. The mismatch between the two lattice structures causes strain hardening (also called precipitation age hardening) of the part (fig. 2.2b). Finally, some metals form different stable phases at different temperatures. Some phases formed by heat treatment, usually by rapid cooling, are hard materials and strengthen the part when formed (see fig. 1.18, martensite by phase transformation of quenched gamma iron).

A small selection of metallic alloys, with typical applications indicating the strengthening mechanism used, is presented in table 2.1. In engineering design, the hardening mechanisms are often used in combination, producing valuable properties. However, with increased strength, the

Table 2.1. | Typical applications and strengthening mechanisms used for a selection of metals and alloys.

Metal or alloy	Typical application	Work hardening	Solution hardening	Precipitation hardening
Pure Al	Kitchen foil	+++		
Pure Cu	Wire	+++		
Cast Al, Mg	Automotive parts		+++	+
Bronze (Cu-Sn), Brass (Cu-Zn)	Marine components	++	+++	+
Non-heat-treatable wrought Al	Ships, cans, structures	+++	+++	
Heat-treatable wrought Al	Aircraft, structures	+	+	+++
Low-carbon steels	Car bodies, structures, ships, cans	+++	+++	
Low-alloy steels	Tools, automotive parts	+	+	+++
Stainless steels	Pressure vessels	+++	+++	+
Cast Ni alloys	Jet engine turbines		+++	+++

Source: Data from Ashby, Shercliff, and Cebon 2007, 134.
Note: +++ routinely used; ++ often used; + sometimes used.

elongation is reduced, so that the material cannot undergo much deformation without fracture.

Strength Testing

In order to evaluate metallic alloys with respect to their mechanical properties, a number of standard tests and specifications have been developed. Information on strength and ductility may be useful in comparisons of materials, characterization of alloys, quality control, and design (Callister 2015c).

Tensile Tests

The strength of a material may be measured in terms of either the stress necessary to cause appreciable plastic deformation or the maximum stress that the material can withstand (Davis 2004). These values can be inferred from tensile tests, where a carefully prepared specimen is deformed with a gradually increasing tensile load, usually up to its fracture, while measuring the resulting deformation confined to a gauge length in the central region.

The output of such a tensile test is recorded on a graph as stress (applied load divided by the original area of the specimen cross section) versus strain (average linear strain, obtained by dividing the elongation of the gauge length by its original length). Figure 2.3 shows a typical stress-strain curve for an alloy. In fact, the shape and magnitude of the stress-strain curve of a metallic alloy will depend on its composition, heat treatment, and prior history of plastic deformation.

Initially, the application of a relatively low load corresponds to moving from the origin up and along the straight line (fig. 2.3). In this region, for most metals, stress and strain are proportional to each other, and the deformation is designated elastic—that is, it is reversible: when the applied load is released, the part returns to its original size (or length). The slope of the linear portion of the curve indicates the material's resistance to elastic deformation, or stiffness, and is called modulus of elasticity, E. The greater the modulus, the stiffer the material and the smaller the elastic strain upon application of a given stress. The modulus of elasticity is expressed in pascals (1 Pa = 1 Newton/m^2) and its multiple units; typical values are given in table 2.2 for selected metals and alloys.

As the material is further loaded, there is a point beyond which the stress is no longer proportional to strain, and plastic, or permanent, deformation occurs (fig. 2.3, point Y). The plot's departure from linearity corresponds to the beginning of plastic deformation (the motion of dislocations resulting in slip of atomic planes over one another), and the stress level at which it occurs is defined as *yield strength* (σ_Y). The magnitude of the yield strength for a metal is a measure of its resistance to plastic deformation.

After yielding, the stress necessary to continue plastic deformation in metals increases to a maximum and then decreases to the eventual fracture (fig 2.3, point T). The *tensile strength* (σ_T) is the stress at the maximum on the

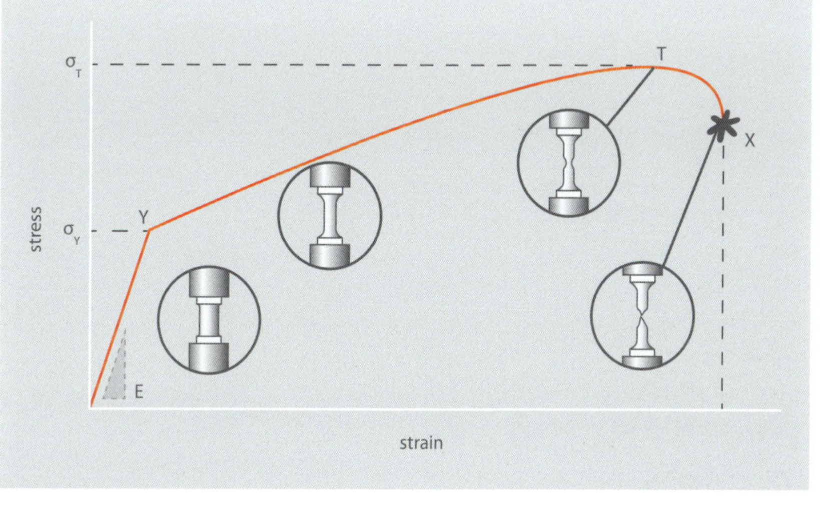

Figure 2.3 | Typical engineering stress-strain behavior (E = modulus of elasticity; σ_Y = yield strength; σ_T = tensile strength; X = fracture point). The circular insets represent the evolution of the specimen geometry during the test (adapted from Callister 2015c, fig. 7.11).

Table 2.2. | Modulus of elasticity for various metal alloys at room temperature.

Metal alloy	Modulus of elasticity (GPa)
Tungsten	407
Steel	207
Nickel	207
Copper	110
Titanium	107
Brass	97
Aluminum	69
Magnesium	45

Source: Data from Callister 2015c, 233.

curve and corresponds to the maximum stress that can be sustained by a structure under tension. Up to this point, the deformation is uniform throughout the specimen. From this maximum, a small constriction, or neck, begins to form, and all subsequent deformation is confined to this region, where fracture ultimately occurs.

Two other useful properties can be inferred from stress-strain plots: ductility and toughness. *Ductility* is a measure of the degree of plastic deformation that has been sustained at fracture. It is determined after rupture and can be expressed quantitatively as either percent of elongation or percent reduction in area. A material that undergoes very little or no plastic deformation upon fracture is termed *brittle*. The tensile behaviors for both ductile and brittle (or fragile) materials are illustrated in figure 2.4.

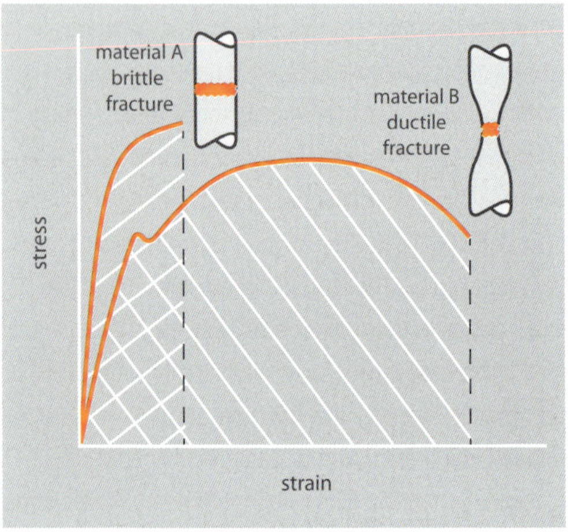

Figure 2.4 | Stress-strain curves for brittle and ductile materials loaded to fracture. The area under each curve represents the energy absorbed by the material prior to failure and corresponds to the material toughness.

Table 2.3. | Mechanical properties for various metal alloys at room temperature.

Metal alloy	Yield strength σ_Y (MPa)	Tensile strength σ_T (MPa)	Ductility (% El in 50 mm)
Molybdenum	565	655	35
Titanium	450	520	25
Steel (1020)	180	380	25
Nickel	138	480	40
Iron	130	262	45
Brass (70Cu-30Zn)	75	300	68
Copper	69	200	45
Aluminum	35	90	40

Source: Data from Callister 2015c, 234.

The *toughness* of a material is its ability to absorb energy in the plastic range. It may be considered to be the total area under the stress-strain curve, which is an indication of the amount of work per volume unit that can be done on the material without causing it to break. In figure 2.4, material B (the larger area under the curve) is tougher than material A. Toughness is a parameter that comprises both strength and ductility.

A general and comparative view of three important mechanical properties—yield strength, tensile strength, and ductility—as measured by tension tests for various materials is presented in table 2.3.

The longstanding practice of using the tensile strength to determine the strength of materials made tensile strength a widely familiar property, and as such it is very useful for the identification of a material, in the same sense that the chemical composition serves to identify a metal or alloy. Furthermore, because the tensile strength is easy to determine and is a quite reproducible property, it is useful for the purposes of specifications and for quality control of a product. An example of the use in conservation is presented in figure 2.5. For the restoration of a severely vandalized bronze sculpture, *The Thinker* by Auguste Rodin, it was necessary to shape the deformation back to its original position. In order to determine whether such additional plastic deformation would pose too great a risk of fracture, test rods were cast in the same alloy as the damaged bronze and subjected to tensile strength tests. Results showed that although the alloy was brittle, controlled plastic deformation would be possible to some degree.

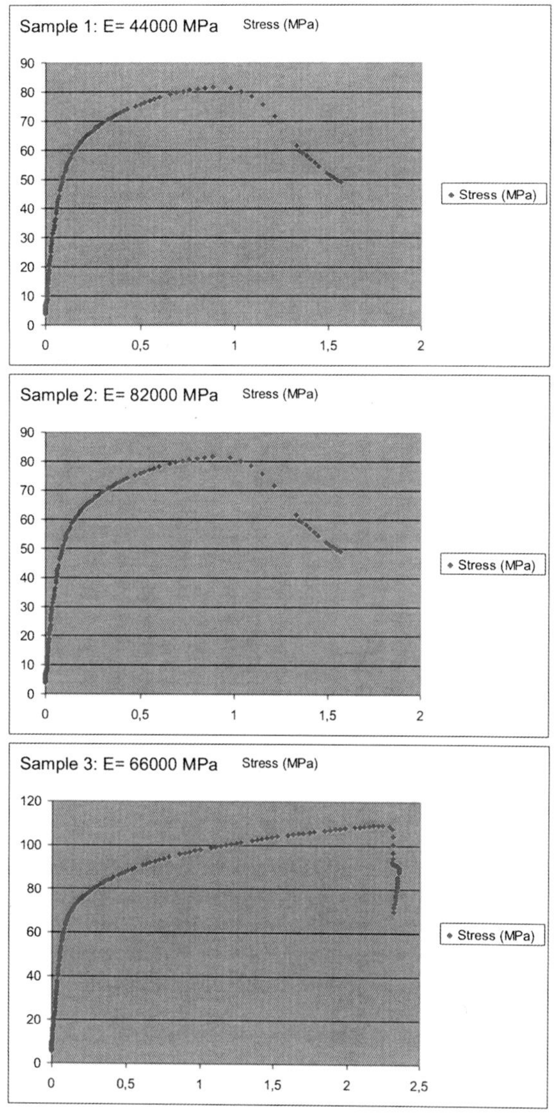

Figure 2.5 | Tensile strength test results for three test rods of the same alloy as the bronze sculpture *The Thinker* by Rodin (reproduced from Bentjees, Davidowitz, and van der Molen 2011, fig. 8). Image: © Corus.

Hardness Tests

In the case of metallic alloys, hardness is usually defined as resistance to indentation. Hardness gives a general indication of the strength of the material and its resistance to scratch and wear (Reardon 2011a).

During the test, an indenter—made of a very hard material such as diamond, hardened steel, or tungsten carbide—is forced into the surface to be tested under controlled conditions of load and rate of application (fig. 2.6a). The indenter is then removed, leaving a notch on the surface of the test piece (fig. 2.6b). The measure of the depth or size of the notch is related to a hardness number. The softer the material, the larger and deeper the indentation and the lower the hardness index number.

Several test methods, using different indenter materials and shapes, have been developed to measure the hardness of materials (table 2.4). The most commonly used are the Brinell test (HB) and the Vickers test (HV), both presenting a broad scale of measurement. There is also the Rockwell test, which offers greater precision within a more limited scale (HRA, HRB, HRC, etc.).

Another method is the Knoop test (HK), which in fact is a microhardness test, because of the elongated shape of the indenter and the light loads applied. This test is therefore suitable for very small or very thin specimens and for measuring the hardness of individual grains and microstructural constituents in a metal alloy. Thin metallic coatings can also be evaluated by performing the Knoop test on the cross section of an embedded sample.

Since the hardness number depends on the applied load, test results should always cite the load that is applied. Measured hardness values are only relative, and care should be exercised when comparing values determined by different techniques. However, values obtained from different hardness methods can be interrelated and converted, also offering an estimate of the equivalent yield strength of a given material (fig. 2.7).

In a study, microhardness measurements were used to characterize piano wires from an 1812 French Érard piano belonging to the Musée de la Musique in Paris. Small samples (about 5 mm long) were prepared for metallography and microhardness testing. The thicker copper wires have a hardness of 150 HV100, while the intermediate brass wires present a mean value of 200 HV100 and the thinnest iron alloy wires, 280 HV100. Such values are in agreement with those found in earlier studies (Goodway and Odell 1987; Goodway 1999), confirming that, in general, the hardness, and therefore the strength, increased as the wire diameter decreased.

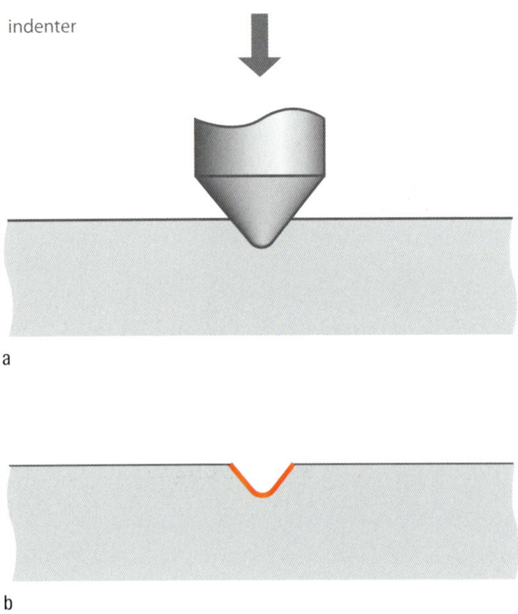

Figure 2.6 | Hardness test: The indenter is applied to the surface under load (a); the depth or size of the indentation (b) is measured and related to a hardness number.

Table 2.4. | General characteristics of hardness-testing methods.

Test		Indenter	Indentation shape	Load (kg)
Brinell		Steel or tungsten carbide (WC) sphere	●	500, 1500, 3000
Vickers		Diamond pyramid	■	1–120
Knoop		Diamond pyramid	◆	0.025–5
Rockwell	A, C, D	Diamond cone	●	60, 100, 150
	B, F, G, E	Steel ball		

Source: Data from Kalpakjian and Schmid 2014, 69.

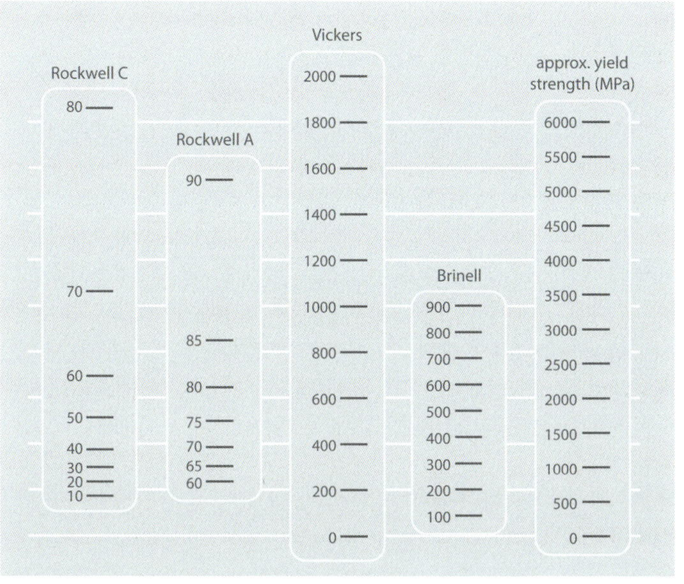

Figure 2.7 | Chart for converting various hardness scales, with comparison to yield strength (adapted from Ashby, Shercliff, and Cebon 2007, fig. 6.5).

DETERIORATIVE PROPERTIES

Corrosion is defined as the deterioration of a material, usually a metal, that results from a reaction with its environment (Jacobson 2017). With a few exceptions, metals are not found in the metallic form in nature but found as ores—that is, an oxide, a sulfide, or some other compound. Metallurgical processes require considerable amounts of energy to extract pure metal from its ore. As a consequence, in their metallic (reduced) form, metals are at a higher, less stable energy level. In certain environments, these metals combine chemically with elements to form compounds and return to their oxidized state (fig. 2.8).

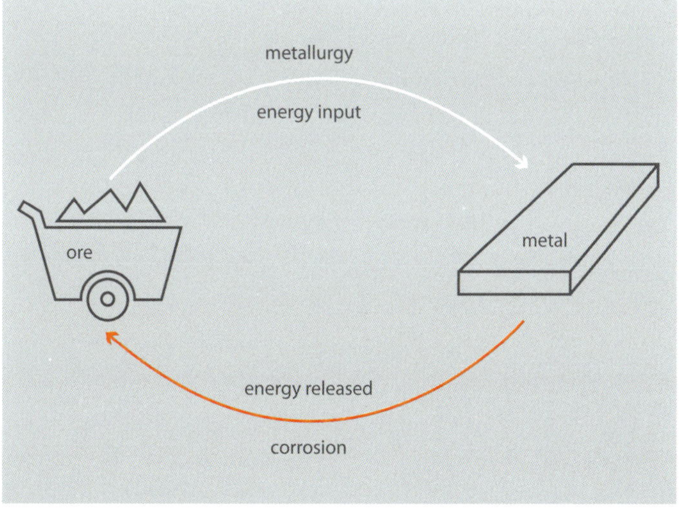

Figure 2.8 | The cycle of corrosion.

Electrochemical Reactions

The corrosion process ordinarily begins at the surface and is normally an electrochemical reaction, in which there is transfer of electrons from one chemical species to another via the metallic conductor. For metallic corrosion to occur, a system in which the electrons liberated within the metal during the oxidation process can be consumed elsewhere at the surface is required. This happens in another process, the reduction reaction. An overall electrochemical reaction must consist of at least one oxidation and one reduction reaction (individually termed half reactions). Stop either half of the oxidation and reduction process, and the corrosion halts. Accelerate either half of the process, and the corrosion rate increases.

If a piece of iron is placed in an acidic solution, a vigorous reaction occurs: iron dissolves, and hydrogen gas is evolved. Iron is converted from its elemental form (Fe^0) into an ion (Fe^{2+}), liberating two electrons in the process. The reaction is represented in figure 2.9a. The metal is then said to be *oxidized*, and the site where this reaction takes place is called the *anode*. The electrons generated in the oxidation reaction migrate inside the metal to an adjacent location, where they take part in the reduction reaction. At this site, hydrogen ions from the acidic solution accept the electrons to become atomic hydrogen, which immediately combines with another hydrogen atom to form molecular H_2. This reduction reaction takes place at the *cathode* site and is responsible for the gas evolution. During metallic corrosion, the oxidation rate is often increased because several different cathodic reactions can take place simultaneously. In acidic aerated environments, for instance, two cathodic reactions are possible—oxygen reduction and hydrogen evolution—increasing the electron demand at the surface and thus increasing the metal dissolution rate (fig 2.9b).

The same accelerator effect can be observed if oxidizing impurities are present, such as rust. In this case a third cathodic reaction—ferric ion reduction—will add to the first two and increase even further the metal dissolution rate.

Factors Affecting Corrosion

Corrosion resistance is not an intrinsic material property. It is the result of the interaction between a metal and a particular environment. Both aspects, together with the nature of the

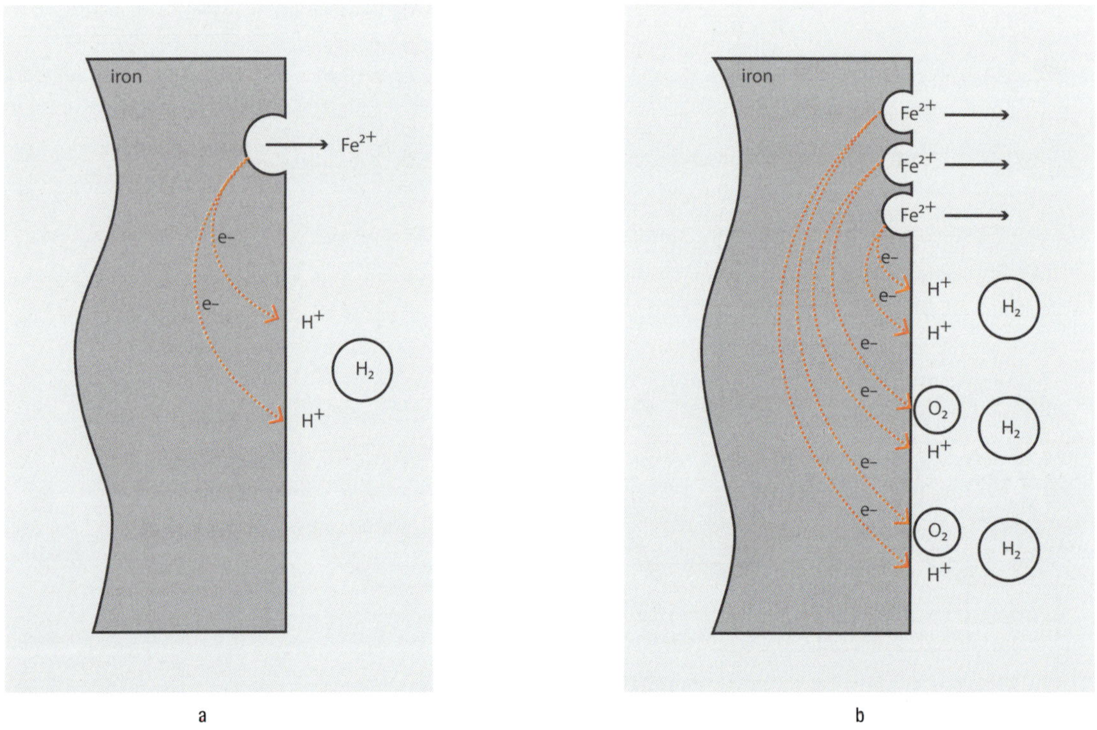

Figure 2.9 | Electrochemical reactions associated with the corrosion of iron in acidic solution, showing the separation of anodic and cathodic sites (a); and an equivalent process in acidic aerated solution involving two cathodic reactions (b) (adapted from Fontana 1986a, figs. 2-3, 2-4).

reaction products, have a pronounced effect on the occurrence, the rate, and the form in which corrosion takes place. These three aspects—metal, environment, and reaction products—are addressed below (Davis 2000a).

Metal

Not all metals oxidize to form ions with the same degree of ease. Metallic materials may be rated with reference to their tendency to undergo corrosion in solutions of their respective ions. Such a tendency is measured as a potential difference between the metal and a reference electrode (fig. 2.10). There are various types of reference electrodes, and all of them can be referenced to the standard hydrogen electrode (SHE), whose potential is taken as zero by internationally accepted convention (IUPAC 2014, 1436).

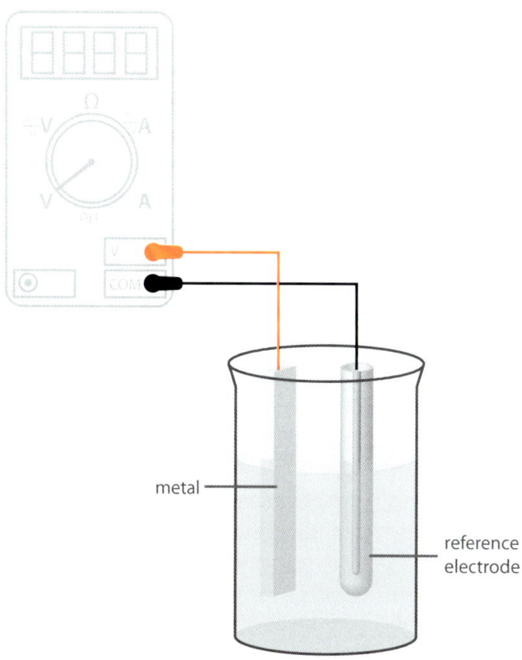

Figure 2.10 | Potential measurement setup: The reference electrode is connected to the "COM" terminal of the millivoltmeter.

Measured potential values are displayed in tables called either electromotive force (emf) series or galvanic series, depending on the conditions in which measurements have been performed. In the emf series, the measured values refer to pure metal electrodes immersed in a solution of their own ions at unit activity and 25°C. The most noble metals are usually at the top of the series and have the highest positive standard electrode potentials. The most active, or less stable, metals are at the bottom of the series and have the most negative standard electrode potentials (table 2.5).

The emf series has limited utility, since the values have been determined for a special set of standard parameters without considering the real conditions. A more realistic and practical ranking is provided by the galvanic series, a listing of the inherent reactivity of different metallic alloys in a specific environment. The galvanic series takes into account all the factors that affect the electrochemical potential of these alloys in a given environment, including alloy composition, heat treatment, surface preparation (mill scale, coatings), surface oxides, environmental composition, and so on. Table 2.6 presents a galvanic series of several commercial metals and alloys in seawater. The galvanic series is quite useful in determining the likelihood of detrimental corrosion effects when dissimilar metals are coupled. The farther apart two metals are in the galvanic series, the more likely that detrimental galvanic corrosion will occur. When two metals are near one another in the galvanic series, the detrimental galvanic effect is lessened.

Table 2.5. | The standard emf series.

	Electrode reaction	Standard electrode potential E (V)
↑	$Au^{3+} + 3e^- \rightarrow Au$	+1.420
	$Pt^{2+} + 2e^- \rightarrow Pt$	~+1.2
Increasingly inert	$Ag^+ + e^- \rightarrow Ag$	+0.800
(cathodic)	$Cu^{2+} + 2e^- \rightarrow Cu$	+0.340
	$2H^+ + 2e^- \rightarrow H2$	0.000
Increasingly active	$Sn^{2+} + 2e^- \rightarrow Sn$	−0.136
(anodic)	$Ni^{2+} + 2e^- \rightarrow Ni$	−0.250
	$Fe^{2+} + 2e^- \rightarrow Fe$	−0.440
↓	$Zn^{2+} + 2e^- \rightarrow Zn$	−0.763
	$Al^{3+} + 3e^- \rightarrow Al$	−1.662

Source: Data from Callister 2015b, 717.

Table 2.6. | The galvanic series in seawater.

↑	Platinum
	Gold
Increasingly inert (cathodic)	Titanium
	Silver
	Stainless steel (passive)
	Nickel (passive)
	Bronzes (Cu-Sn alloys)
	Copper
	Brasses (Cu-Zn alloys)
	Nickel (active)
	Tin
	Lead
	Stainless steel (active)
	Cast iron
	Iron and steel
	Aluminum alloys
Increasingly active (anodic)	Cadmium
	Commercially pure aluminum
	Zinc
↓	Magnesium and magnesium alloys

Source: Data from Callister 2015b, 720.

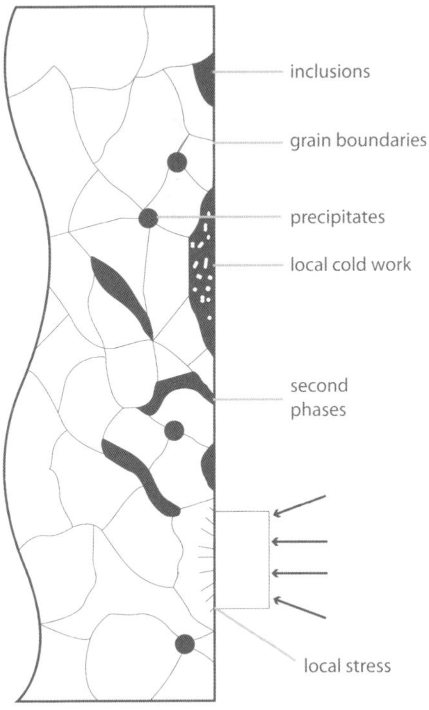

Figure 2.11 | Metallurgical factors affecting corrosion (adapted from Budinski and Budinski 2010, fig. 6-6).

When consulting the galvanic series, recall that it is a ranking of materials for a specific environment under a specific set of conditions. Application of a particular series to a different set of conditions or to a different environment can lead to erroneous prediction.

Besides the composition, there are many other metallurgical factors that are relevant to the chemical stability of the alloys. Microstructural features have an important effect on the corrosion resistance because in highly deformed metals, the grain structure is completely disrupted, which means that the material is somewhat more reactive in a number of environments. In commercial alloys, many metallurgical factors—such as chemical segregation, the presence of multiples phases, inclusions, and nonuniform stresses—constitute particular sites inside the alloy. When exposed to an aggressive environment, such heterogeneities will react differently from the matrix, and a potential difference will eventually establish between both, increasing the corrosion rate and often leading to a selective form of corrosion (fig. 2.11).

Environment

With regard to the environment, the essential requirement for electrochemical corrosion to take place is the presence of an electrolyte—a solution that conducts electricity by migration of ions—on the surface of the metal. Important characteristics of the aqueous solution, such as conductivity, acidity or alkalinity, oxidizing power, and solubility, in combination with the characteristics of the metal, will determine the corrosion behavior of a metal-environment system (Davis 2000c).

In atmospheric corrosion, the electrolyte comes from the humidity in air. The most important factor is moisture—rain, dew and condensation, or high relative humidity (RH)—overriding pollution or the lack of it. In the absence of moisture, most contaminants would have little or no corrosive effect. The danger of corrosion appears to rise rapidly with increasing RH, but a critical RH value, at which corrosion rates suddenly increase, seems to be a function of the contaminants present and of the hygroscopic properties of the corrosion products (Leygraf and Graedel 2000). Atmospheres are traditionally classified into four different types: industrial, marine, rural, and indoor. Industrial atmosphere is characterized by pollution composed mainly of sulfur compounds, acid chlorides, phosphates, and nitrates that produce a highly corrosive wet acid film on exposed surfaces. A marine atmosphere is laden with fine particles of sea salt carried by the wind and settling on exposed surfaces. Rural atmospheres do not contain strong chemical contaminants but have some organic and inorganic dusts. Their principal corrosive constituents are moisture and gaseous elements such as oxygen and carbon dioxide. Moisture is particularly important in tropical climates, producing a rather corrosive environment. The indoor atmosphere, while ordinarily considered mild, can sometimes in fact be very severe; even if the absolute pollutant concentration is not high, it is contained in a limited and confined space (Hatchfield 2002).

Regarding corrosion by water, for the most part, the common structural metals possess adequate resistance to high-purity water, but contaminants and dissolved oxygen make water corrosive. Contaminants often found in water that contribute to corrosion include chloride salts, sulfides from wells, sewage, and carbon dioxide from combustion products in the air. Many other man-made contaminants can be found in local areas where industries discharge their waste into water streams.

The corrosiveness of soils can be estimated from their electrical resistivity. Low values correspond to permeability to water and/or the presence of dissolved species, both of which contribute to making the soil corrosive. Substances dissolved in the moisture contained in soils are very numerous; they range from traces to very substantial amounts in some cases, causing widely different behaviors with regard to corrosion (Roberge 2007, 103)

Reaction Products

Under particular environmental conditions, certain metals and alloys react to form a thin film that enhances their corrosion resistance by acting as a barrier between the surface and the corrosive agent. This phenomenon, called *passivity*, refers to the loss of chemical reactivity undergone by certain metals and alloys under particular conditions. In order to act as a barrier, the passive film has to be continuously adherent to the metallic surface, and it has to be chemically stable and impermeable in the environment. The emf series (see table 2.5) shows that aluminum is an extremely reactive metal.

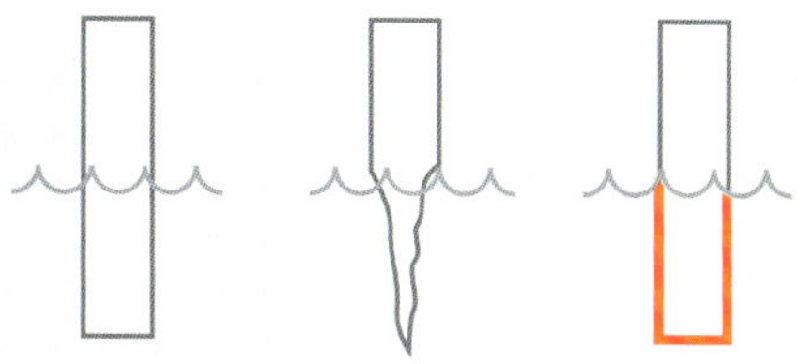

Figure 2.12 | Three behaviors of metal in the environment (adapted from Davis 2000a, fig. 1).

However, because of the formation of a very adherent and protective oxide layer under normal atmospheric conditions, aluminum can be used for many architectural and structural purposes. Other metals also present insoluble surface products, but if these products are porous, they will provide little or no protection. It is important to know which products are the most stable and what degree of protection is provided by the solid reaction products.

Thus, depending on the nature of the products formed during reaction with the environment, the metal will be classified as inert (no reaction at all), active (there is reaction, and the products are soluble, allowing further corrosion), or passive (there is reaction, but the products are insoluble and may precipitate on the surface with a certain or possible barrier effect) (fig. 2.12).

When considering the active-passive behavior of several alloys, it can be seen that several important shifts are introduced in the galvanic series previously presented. The principal reason for this is that some metals and alloys develop stable protective corrosion product films when they interact with the environment. When such a film forms, an alloy is referred to as being in the passive condition. Stainless steel is an example of this class of material. Table 2.6 shows the large shift in position for an 18-8 steel in the active (corroding) condition to a much more noble position in the passive condition.

The active-passive behavior can be encompassed by potential-pH, or Pourbaix, diagrams, which consider the possibilities of reaction between metal and water under standard conditions. They make a correlation between the many stability zones for the different compounds and the thermodynamic probability of corrosion, providing a useful means of summarizing the thermodynamic behavior of a metal and associated species in given environmental conditions.

Potential-pH diagrams are typically plotted for various equilibria on normal Cartesian coordinates with potential (E) as the ordinate, and pH value as the abscissa. Changes in other variables, such as oxygen concentration, tend to be reflected in changes in the corrosion potential. Simplified diagrams for several metals in water at 25°C are shown in figure 2.13, which presents the different regions corresponding to the three possible behaviors of immunity, corrosion and passivity.

Potential-pH diagrams can be used to design methods of corrosion control. For instance, for a given metal, the conditions in which soluble ions form (corrosion, point A, fig. 2.14) must be avoided, and for the same metal, environmental conditions must be chosen (by changing pH and oxidizing capacity) so that the location falls into a region of passivity (insoluble products, points B and B', fig. 2.14) or immunity (metal thermodynamically stable, point C, fig. 2.14). Changing the potential to the immunity region—by applying a cathodic current or coupling a sacrificial anode—is the principle of cathodic protection. In the passive region, the corrosion resistance of the material relies on the integrity of the protective film. In the immunity region, however, the corrosion resistance is imparted by the thermodynamic stability of the metal itself, which is much more reliable.

There are limitations to the use of potential-pH diagrams, because they are the result of thermodynamic calculations and do not provide any information about reaction rates (Fontana 1986a, 452). Moreover, diagrams are based on the reaction of pure metal with the constituents of water. This approach neglects all the possibly important effects of other ions present in solution (chloride, nitrate, reduced sulfur species, etc.). Also, they are calculated for a given temperature, usually 25°C, and temperature changes can greatly modify the size of the stability fields. In addition, diagrams assume that all insoluble products result in passivity, which is not always the case in practice, because many insoluble products are not continuous or are loosely attached to the metal surface, providing limited protection against further corrosion.

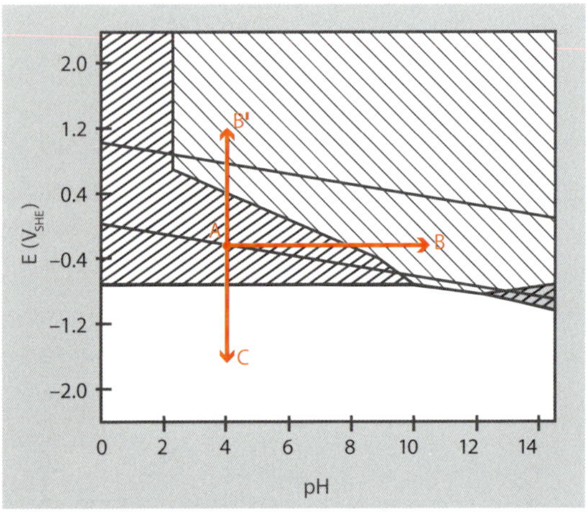

Figure 2.14 | Potential-pH diagram for iron in the presence of water at 25°C (adapted from Pourbaix 1973, fig. 86). Changes in potential and pH can shift the condition from corrosion (point A) to passivity (points B and B') or immunity (point C).

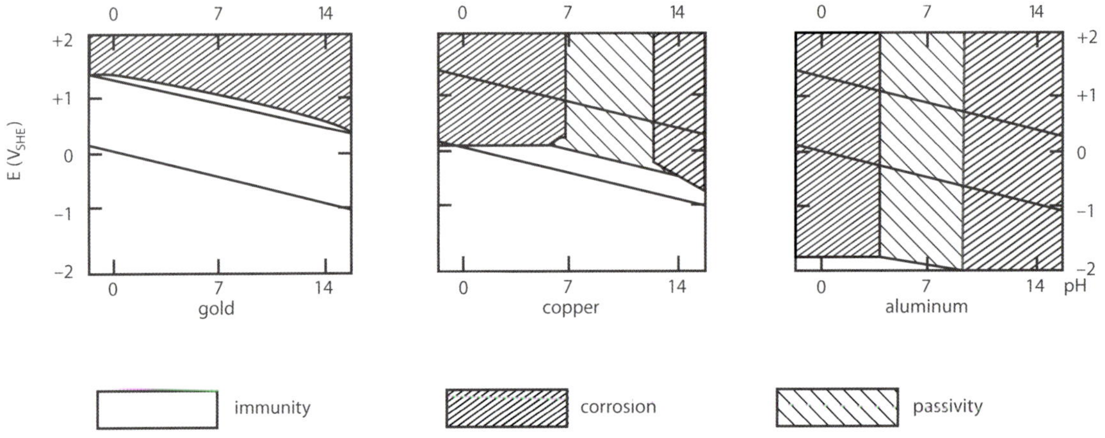

Figure 2.13 | Potential-pH diagram for three metals in water (adapted from Pourbaix 1973, fig. 55).

Forms of Corrosion

Corrosion of metallic alloys can take different forms, which can often be identified by visual observation (Elliott 2005). Recognition of the type of corrosive attack helps in determining its cause, measuring the extent of attack, and recommending or developing methods of control. Nevertheless, the combination of metals used in a system and the wide range of environments encountered often cause more than one type of attack to occur. Some of the more usual corrosion forms (Fontana 1986b) are described below.

Uniform Attack

Uniform attack is a very common form of corrosion, often characterized by the presence of scale or deposit on the surface. As the name implies, the electrochemical reactions proceed uniformly over the surface, alternating cathodic and anodic sites randomly. As a result, the metal becomes thinner and eventually fails (fig. 2.15). Typical examples include the rusting of steel, the green patina formation on copper, and the tarnishing of silver.

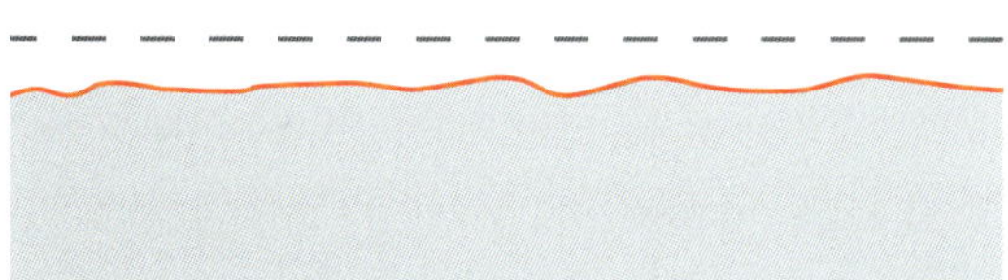

Figure 2.15 | Uniform corrosion attack.

Galvanic Corrosion

Galvanic corrosion is a corrosion process that occurs when two different metals or alloys are electrically connected and immersed in a conductive solution, and a potential difference is established between the two metals, producing an electron flow between them. The less noble metal or alloy in the environment becomes the anode and will undergo corrosion, while the more noble becomes the cathode and corrodes very little or not at all (fig. 2.16).

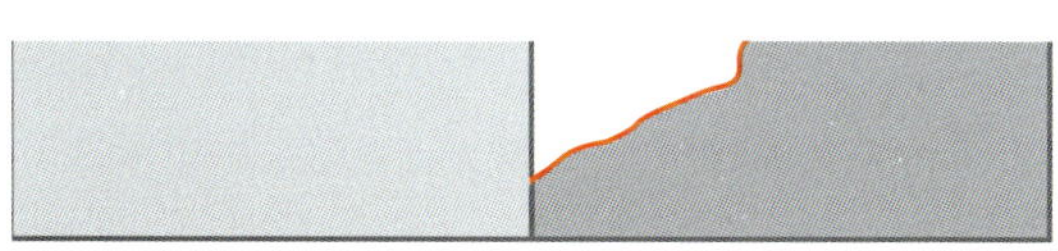

Figure 2.16 | Galvanic corrosion.

A prediction of galvanic relationships is listed in the galvanic series in the appropriate environment. For instance, table 2.6 presents the galvanic series for metals in seawater at room temperature. The rate of galvanic attack depends on the relative position of both alloys in the series and on their relative sizes (surface area exposed to the solution). Dissolution of the more active metal will in fact be accelerated for metals situated far apart in the series as well as with increasing ratio of cathode-to-anode areas.

Galvanic corrosion does not occur when metals are close in the galvanic series, nor does it occur in a dry environment, where there is no electrolyte to carry charge. When adequate conditions cannot be found, dissimilar metals must be insulated from each other with a third, nonmetallic material that does not absorb or retain humidity (ASTM International 2014).

Crevice Corrosion

Crevice corrosion is a form of localized attack that occurs at narrow openings or spaces between a metal and another material, metallic or not. This attack results from a difference in aeration between the electrolyte within the crevice, which is depleted in oxygen, and the electrolyte outside the crevice, where oxygen is present. The result is current flowing between the oxygen-rich part outside the crevice (cathode) and the adjacent oxygen-depleted zone in the crevice (anode). This is sometimes called a *concentration cell*, or a *differential aeration cell*. Once the crevice attack has initiated, the anodic area becomes acidic, and this change further accelerates local cell action. Moreover, such a localized environment often contains halide anions (e.g., chlorides) and can become much more aggressive than a bulk solution. Crevices occur at gaskets, threaded joints, riveted seams, and lap joints; beneath coatings (filiform corrosion, discussed below); and at any location where close-fitting surfaces are present.

This type of attack also occurs under naturally formed (non-manmade) deposits and is referred to as *underdeposit corrosion* (fig. 2.17a). Deposits include water-borne precipitates, transported particles, corrosion products, biological materials, and a variety of contaminants such as grease, sand, and mud. They can be innately aggressive (e.g., deposits containing corrosive substances such as sulfur-containing or chloride-containing species) or simply serve as an occluding medium beneath which concentration

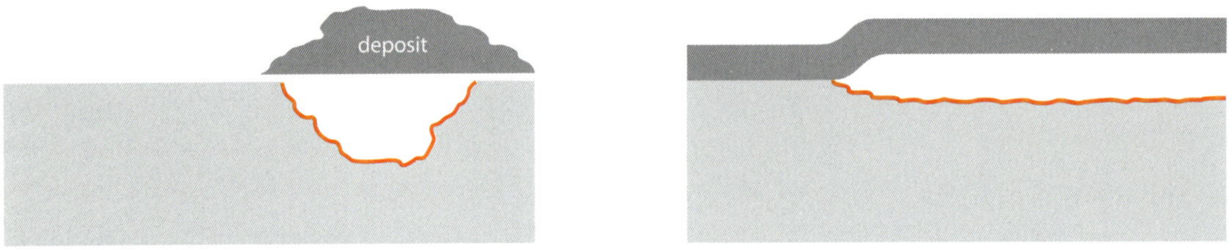

a

b

Figure 2.17 | Crevice corrosion: Underdeposit (a); and filiform corrosion (b).

cells develop. Severe wastage from pitting can occur under these deposits.

A special type of crevice corrosion is *filiform corrosion* (fig. 2.17b). It occurs under painted or plated surfaces when moisture permeates the coating. The source of initiation is usually a defect or scratch in the coating, and the corrosion evolves as filaments—that is, fine tunnels composed of corrosion products underneath the cracked coating. Under transparent coatings, the attack appears as a network of corrosion-product trails. It does not weaken or destroy metallic components but irreversibly affects the surface appearance. The most important environmental variable is the relative humidity (RH) of the atmosphere, occurring primarily between 65% and 90% RH. Where a coating is required, it should exhibit low water vapor transmission characteristics and excellent adhesion, enabled by careful surface preparation prior to coating. Filiform corrosion has been observed in coated steel cans, coated tin-plated steel, aluminum foil–laminated packaging, painted aluminum and magnesium, and other lacquered metallic items placed in areas with high humidity.

Pitting

Pitting is a form of very localized and intense attack that results in small pits or holes progressing deep into the metal. Pitting occurs when a localized area of a metallic surface becomes anodic with respect to the rest of the surface (fig. 2.18). Causes of pitting are local heterogeneity on the metal surface, local loss of passivity, mechanical or chemical rupture of a protective coating, and so on. It is observed especially in alloys presenting a protective passive film (stainless steel, titanium and aluminum alloys) that has been disrupted locally by internal causes (e.g., slight variation in composition) or external causes (e.g., scratching). Pits are difficult to detect because of their small size and because they are often covered with corrosion products.

Selective Corrosion

Selective corrosion is a corrosion process whereby one constituent of the metallic alloy is selectively removed, leaving behind a porous layer of the more noble metal (fig. 2.19). As with other types of corrosion, the primary cause of dealloying is the galvanic effect between the different elements that compose the alloy.

The most common example is the dezincification of brass, in which zinc is selectively leached from a copper-zinc base alloy. The mechanical properties of the alloy are significantly impaired, since only a porous mass of copper remains in the region that has been dezincified. A number of alloy systems are susceptible to this process. Examples are the selective leaching of aluminum from aluminum bronzes and the graphitic corrosion of gray iron.

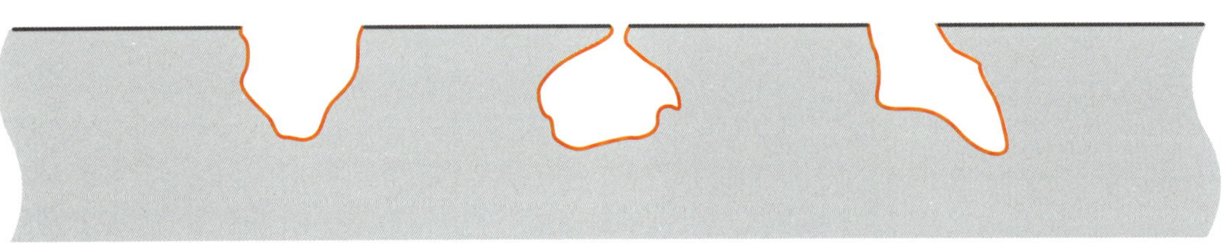

Figure 2.18 | Pitting corrosion.

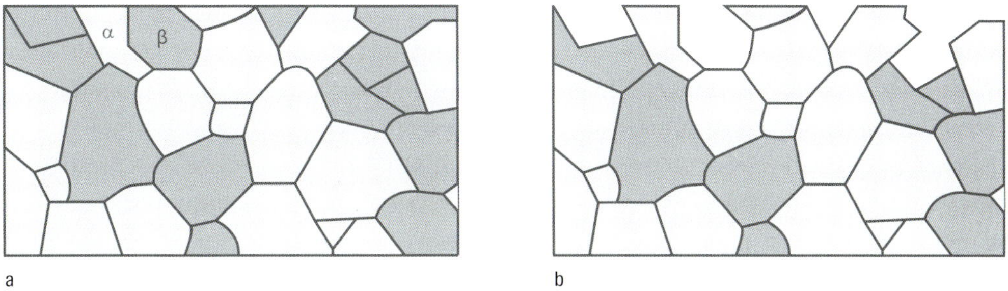

Figure 2.19 | Biphasic metal before (a) and after (b) preferential corrosion of β phase, leaving the surface α-rich.

Intergranular Corrosion

Intergranular corrosion is defined as the selective dissolution of grain boundaries, or closely adjacent regions, without appreciable attack of the grain itself. The driving force is a difference in potential that develops between two zones of different chemical composition: the thin grain boundary, at which heterogeneities concentrate (mismatch in crystal structure, precipitate, or segregation), and the bulk of immediately adjacent grains (fig. 2.20a). As the cracks develop inward, they follow the grain boundaries, with the result that the outer layers of grains become loose. Since most intergranular corrosion is the result of small differences in composition at grain boundaries, the metallurgical history of an alloy is as important as its average chemical composition. This is because heat treatments and cold working affect not only the size and shape of the grains but also the composition, location, amount, and size of the constituents.

A particular form of intergranular corrosion is *exfoliation* (fig. 2.20b). It occurs when the part exposed to the corrosive environment has been previously extensively cold-worked, producing an elongated grain structure parallel to the surface. Exfoliation manifests by the lifting of the surface grains, induced by the expanding corrosion products formed at the grain boundaries just below the surface. It is often observed on extruded sections.

Stress Corrosion

Stress corrosion cracking (SCC) is a cracking phenomenon caused by the simultaneous action of a surface tensile stress and the presence of a specific corrosive environment. During SCC, the metal or alloy is virtually unattacked over most of its surface, while fine cracks progress in a direction perpendicular to the stress that contributes to their initiation and propagation (fig. 2.21). The stress that produces SCC need not be externally applied. It may be a residual stress that results from rapid temperature changes and uneven contraction, or it may be induced in two-phase alloys for which each phase has a different coefficient of thermal expansion.

Corrosion Testing and Evaluation

The corrosion susceptibility of metals in different situations is usually characterized by submitting specimens to particular test conditions and evaluating them after a given period of exposure time. Tests should be carried out in an environment that simulates as closely as possible the one in which the part will be used. Since a large number of factors affects corrosion behavior, there is no universal corrosion test. Methods and standards related to the overall behavior and performance of materials are proposed by organizations such as ASTM International, NACE International, International Organization for Standardization (ISO), European Standard (EN), and Japanese Industrial Standards (JIS).

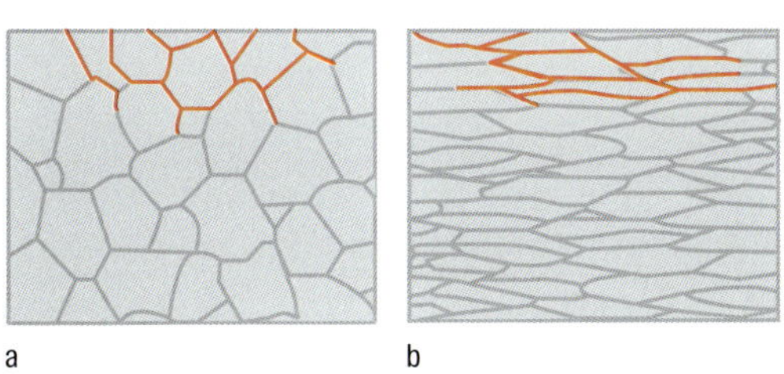

Figure 2.20 | Intergranular corrosion (a); and exfoliation (b).

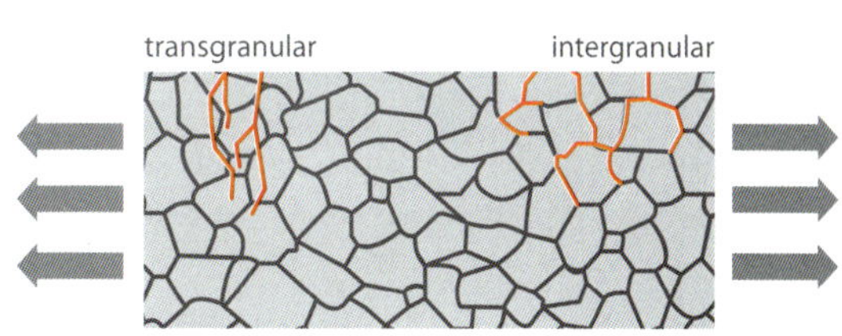

Figure 2.21 | Stress corrosion.

Various types of corrosion-testing methods are used to investigate corrosion failures and to evaluate the resistance to corrosion of metals and alloys for service in specific applications (Baboian 2005). They can be conducted under "natural" or "artificial" conditions. In the first group—called field tests—specimens are exposed for very long periods of time in locations ideally similar to those where the metals will be employed, such as atmospheres (marine, industrial, rural), waters (distilled, fresh, salt, stagnant), and soils (dry, moist, clay) (fig. 2.22). The second group—laboratory tests—is performed under more aggressive and reproducible conditions for shorter periods.

Laboratory tests are very useful for comparing the relative performance of different materials in a well-defined condition, for screening metals and alloys for further testing, and for gaining understanding of corrosion processes. They cannot be used as a guarantee of successful performance, because laboratory conditions never completely mimic nature. Moreover, the acceleration of certain factors during testing may cause other corrosion mechanisms to become active. Thus, the results may not reflect those that would be obtained under actual service conditions. For this reason, results of laboratory tests must be interpreted carefully. They can be related to expected actual service behavior only when close correlation with long-term service results has been established.

Among tests carried out in the laboratory, immersion, cabinet, and electrochemical tests are very common (Davis 2000b). Immersion testing is a frequently used procedure for evaluating the corrosion resistance in aqueous solutions. It provides a preliminary screening of best-suited materials, but the results cannot be extrapolated to predict system lifetime. Tests can be performed by simple immersion—in which specimens of the candidate material are exposed to the test medium for a period of time, and the weight loss is measured—or by alternate immersion, a test procedure in which a specimen is cyclically immersed for a period of time in the environment, then removed and dried before being immersed again. Moreover, in order to simulate as closely as possible the actual corrosion mechanisms, immersion can be combined with the application of stress, simulation of crevices, or galvanic couples. Figure 2.23 shows a device specifically designed to evaluate the effect of leaching rain on the corrosion behavior of patinated outdoor bronze through dropping tests simulating a severe runoff condition.

Cabinet testing involves tests conducted in a closed cabinet where the climatic conditions are controlled and mostly designed to accelerate specific corrosion processes. Though results cannot be directly correlated to the performance in natural atmospheres, they are useful for screening similar products. Among the many types of cabinet testing, salt spray testing, or salt fog testing, is the most widely used (ASTM International 2016) (fig. 2.24). Cabinets are very often employed for testing the performance of organic and inorganic coatings on metals.

Because of the electrochemical nature of corrosion, electrochemical tests—measurements of the electrical properties of the metal-solution interface—are extensively used to study corrosion reactions and mechanisms as well as

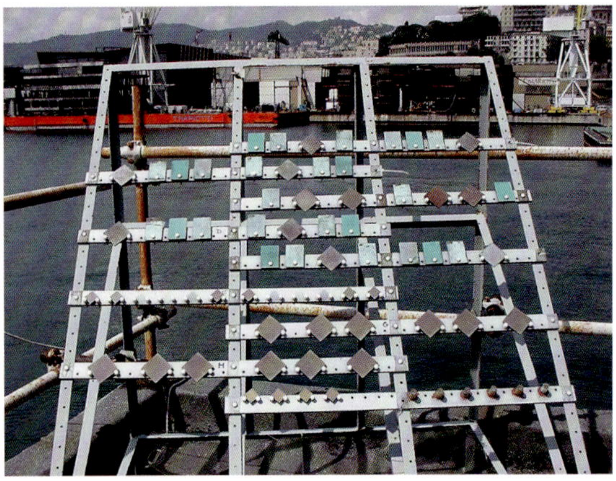

Figure 2.22 | Example of atmospheric corrosion testing station: Bronze samples on a rack in the urban marine environment inside Genoa Harbor (reproduced from Letardi and Golfomitsou 2013, fig. 1).

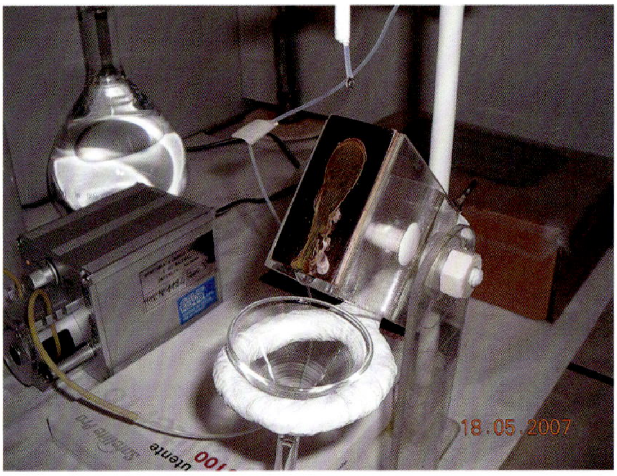

Figure 2.23 | Rain dropping device (reproduced from Bernardi et al. 2009, fig.1).

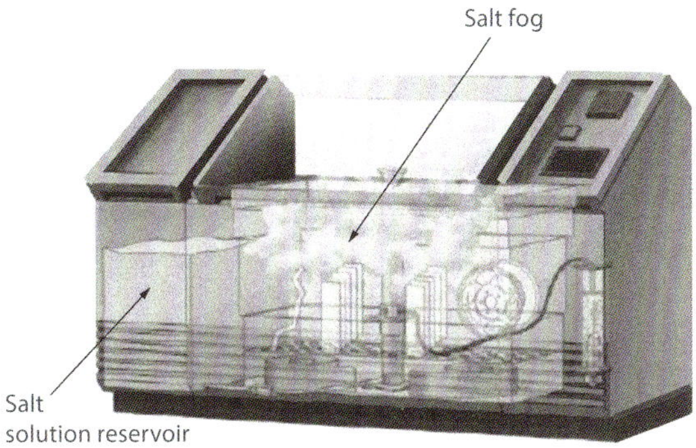

Figure 2.24 | Salt fog test chamber (reproduced from Roberge 2012, fig. 7.44).

to monitor and control corrosion in service. They can directly amplify the impact of corrosion processes, permitting very precise and sensitive measurements to be carried out. In particular, they are used to evaluate the passive-active behavior of alloys, the protectiveness of coatings, and the performance of inhibitors, as well as to estimate corrosion rates (fig. 2.25) (Costa 2003).

Whatever the method, it is essential to examine the test specimens visually in order to identify the form of corrosion and employ the appropriate technique to evaluate and monitor it. For instance, calculating corrosion rates by weight-loss measurements or electrochemical techniques is adequate for metals that corrode uniformly, but it is not adequate in the case of localized corrosion, such as pitting and intergranular attack.

In the case of uniform corrosion, rates are expressed as weight loss (g/m^2 · day), as corrosion current (mA/cm^2), or as penetration rate (mm/y^1, or mpy). Corrosion rates are useful for assessing the corrosive properties of environments, which are classified in different categories, from very low to extremely high corrosion rates. Such classification provides an estimation of the serviceability of different metals and alloys in a given environment (ISO 2012). Starting with a broad classification encompassing a wide range of atmospheres, some categories are further subdivided with the purpose of classifying low-corrosivity indoor atmospheres—for example, places where electronic devices, sophisticated technical products, or works of art and historical objects are stored (ISO 2006).

Estimate of corrosion rates based on weight loss should not be used to evaluate particular forms of corrosion. Localized corrosion, such as crevice and pitting, intergranular corrosion, selective leaching, and stress-corrosion cracking (SCC), act only on a small portion of a component. But

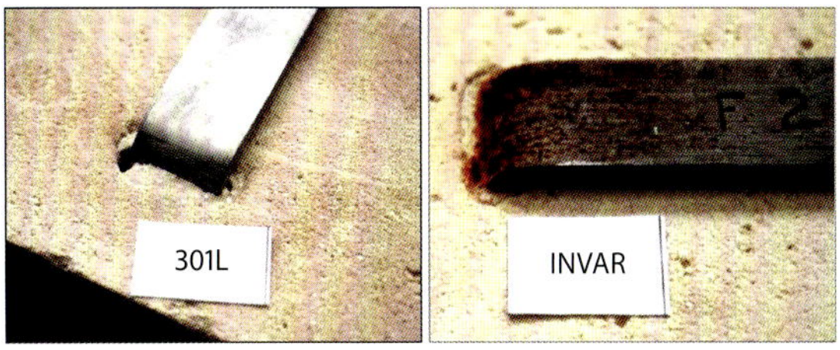

a

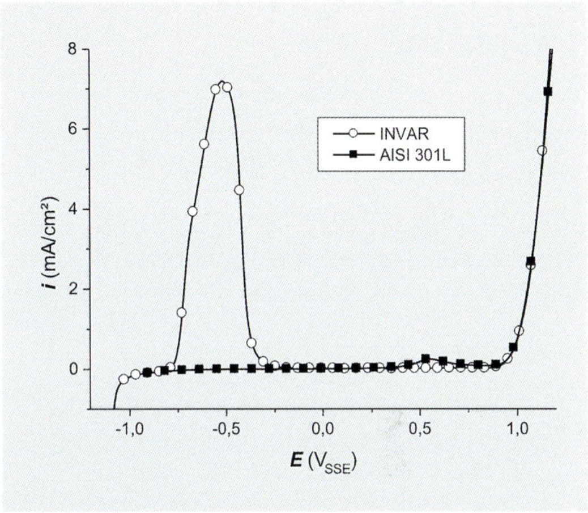

b

Figure 2.25 | Evaluation of compatibility of sealing products with iron alloy clamps, Invar, and stainless steel: After five months' exposure in a climatic chamber, the Invar sample shows visual signs of corrosion in the region in contact with plaster, while the stainless steel clamp does not (a); anodic polarization of both metals in plaster extract as electrolyte (b), carried out in a few hours—an intense current peak can be observed in the case of Invar, attesting to the susceptibility of the alloy in contact with the sealing product (i = current density, SSE = silver sulphate electrode) (reproduced from Costa and Texier 2010, figs. 2b, 3b).

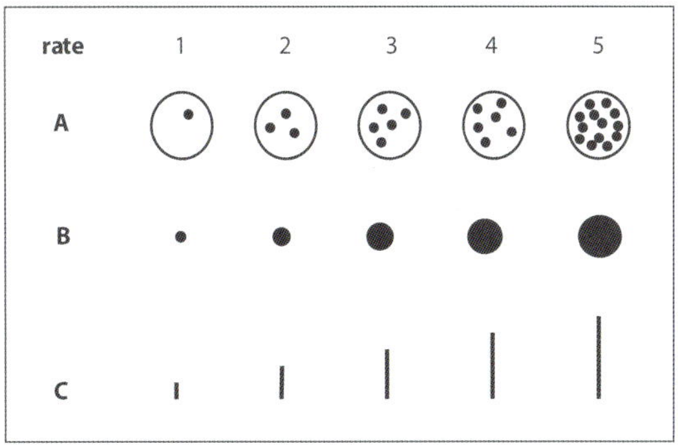

Figure 2.26 | Rating charts for corrosion pits: density (A), ranging from 2.5 x 10^3 to 5 x 10^5/m^2; size (B), ranging from 0.5 to 24.5 mm^2; depth (C), ranging from 0.4 to 6.4 mm (adapted from ASTM International 2013, fig. 2).

localized corrosion occurs at a rate many orders of magnitude greater than that of uniform corrosion. Particular forms of corrosion should be described in detail and reported with reference to the corresponding standards. In the case of pitting corrosion, for example, the corroded surface is usually photographed and compared with standard charts used to rate pits by their density, size, and depth (fig. 2.26). This method offers a simple means for storing data for comparison with other results.

REFERENCES AND BIBLIOGRAPHY

Ashby, M., H. Shercliff, and D. Cebon. 2007. *Materials Engineering, Science, Processing and Design*. Oxford, MA: Elsevier.

ASTM International. 2013. *ASTM G46. Standard Guide for Examination and Evaluation of Pitting Corrosion*. West Conshohocken, PA: ASTM International.

———. 2014. *ASTM G82. Standard Guide for Development and Use of a Galvanic Series for Predicting Galvanic Corrosion Performance*. West Conshohocken, PA: ASTM International.

———. 2016. *ASTM B117. Standard Practice for Operating Salt Spray (Fog) Apparatus*. West Conshohocken, PA: ASTM International.

Baboian, R., ed. 2005. *Corrosion Tests and Standards: Application and Interpretation*. 2nd ed. West Conshohocken, PA: ASTM International.

Beentjes, T., T. Davidowitz, and R. van der Molen. 2011. “Treatment of the Damaged Bronze of Rodin’s the Thinker from the Singer Museum in Laren, The Netherlands: An Innovative Approach.” In *Metal 2010: Proceedings of the Interim Meeting of the ICOM-CC Metal Working Group, October 11–15, Charleston, South Carolina*, edited by P. Markidian, C. Chemello, C. Watters, and P. Hull, 269–77. Clemson, SC: Clemson University.

Bernardi, E., C. Chiavari, B. Lenza, C. Martini, L. Morselli, F. Ospitali, and L. Robbiola. 2009. “The Atmospheric Corrosion of Quaternary Bronzes: The Leaching Action of Acid Rain.” *Corrosion Science* 51:159–70.

Budinski, K. G., and M. K. Budinski. 2010. *Engineering Materials: Properties and Selection*. Upper Saddle River, NJ: Prentice Hall.

Callister, W. D. 2015a. “Deformation and Strengthening Mechanisms.” In *Fundamentals of Materials Science and Engineering*, 280–323. New York: John Wiley and Sons.

———. 2015b. "Corrosion and Degradation of Materials" In *Fundamentals of Materials Science and Engineering*, 711–54. New York: John Wiley and Sons.

———. 2015c, . “Mechanical Properties.” In *Fundamentals of Materials Science and Engineering*, 216–78. New York: John Wiley and Sons.

Costa, V. 2003. “Electrochemistry as a Conservation Tool.” In *Conservation Science 2002: Papers from the Conference Held in Edinburgh, Scotland, 22–24 May 2002*, edited by J. H. Townsend, K. Eremin, and A. Adriaens, 88–95. London: Archetype Publications.

Costa, V., and A. Texier. 2010. “Restoration of Cultural Heritage: Evaluation of the Compatibility between Metals and Sealing Products.” *Journal of Solid State Electrochemistry* 14:403–5.

Davis, J. R., ed. 2000a. “Basic Concepts Important to Corrosion.” In *Corrosion: Understanding the Basics*, 21–48. Materials Park, OH: ASM International.

———. 2000b. “Corrosion Testing and Monitoring.” In *Corrosion: Understanding the Basics*, 427–74. Materials Park, OH: ASM International.

———. 2000c. “Types of Corrosion Environments.” In *Corrosion: Understanding the Basics*, 193–236. Materials Park, OH: ASM International.

———. ed. 2004. *Tensile Testing*. 2nd ed. Materials Park, OH: ASM International.

Elliott, P. 2005. “Gallery of Corrosion Damage.” In *Corrosion: Materials*, 629–46. ASM Handbook. Materials Park, OH: ASM International.

Fontana, M. G. 1986a. *Corrosion Engineering*. New York: McGraw-Hill.

———. 1986b. “Eight Forms of Corrosion.” In *Corrosion Engineering*, 39–152. New York: McGraw-Hill.

Goodway, M. 1999. “The Relation of Hardness to Strength in High-Phosphorus Iron Wire.” *Historical Metallurgy* 33 (2): 104–5.

Goodway, M., and J. S. Odell. 1987. *The Metallurgy of 17th and 18th Century Music Wire*. New York: Pendragon Press.

Hatchfield, P. 2002. *Pollutants in the Museum Environment: Practical Strategies for Problem Solving in Design, Exhibition and Storage*. London: Archetype Publications.

International Union of Pure and Applied Chemistry [IUPAC]. 2014. *Compendium of Chemical Terminology: Gold Book*. Version 2.3.3. Research Triangle Park, NC: IUPAC.

ISO [International Organization for Standardization]. 2006. *ISO 11844. Corrosion of Metals and Alloys—Classification of Low Corrosivity of Indoor Atmospheres*. Part 1, “Determination and Estimation of Indoor Corrosivity”; part 2, “Determination of Corrosion Attack in Indoor Atmospheres”; part 3, “Measurement of Environmental Parameters Affecting Indoor Corrosivity.” Geneva: International Organization for Standardization.

———. 2012. *ISO 9223. Corrosion of Metals and Alloys—Corrosivity of Atmospheres—Classification, Determination and Estimation*. Geneva: International Organization for Standardization.

Jacobson, G. A. 2017. *Corrosion Basics*. NACE International. *www.nace.org/Corrosion-Basics/*.

Kalpakjian, S., and S. R. Schmid. 2014. *Manufacturing Engineering and Technology*. 7th ed. Upper Saddle River, NJ: Pearson.

Letardi, P., and S. Golfomitsou. 2013. “Atmospheric Corrosion of Bronze: A Field Exposure Study of the Correlation of Environmental Data with Patina Growth in an Urban-Marine Environment.” In *Metal 2013: Interim Meeting of the ICOM-CC Metal Working Group, September 16–20, Edinburgh*, edited by E. Hyslop, V. Gonzalez, L. Troalen, and L. Wilson, 153–58. Edinburgh: Historic Scotland.

Leygraf, C., and T. E. Graedel. 2000. *Atmospheric Corrosion*. London: John Wiley and Sons.

Pourbaix, M. 1973. *Lectures on Electrochemical Corrosion*. New York: Plenum Press.

Reardon, A. C. 2011a. “Hardness Testing.” In *Metallurgy for the Non-Metallurgist*, 2nd ed., edited by A. C. Reardon, 446. Materials Park, OH: ASM International.

———. 2011b. “Mechanical Properties and Strengthening Mechanisms.” In *Metallurgy for the Non-Metallurgist*, 2nd ed., edited by A. C. Reardon, 49–72. Materials Park, OH: ASM International.

Roberge, P. R. 2007. *Corrosion Inspection and Monitoring*. Hoboken, NJ: John Wiley and Sons.

———. 2012. *Handbook of Corrosion Engineering*. 2nd ed. New York: McGraw-Hill.

CHAPTER 3 | CHARACTERIZATION

The first and key step in the conservation of an artifact is to characterize it—that is, to know and understand both its immaterial and material aspects. A complete knowledge of materials and techniques used to create the object is also fundamental in defining the artifact's conservation condition and determining whether intervention is needed.

The kind of information necessary for characterization does not come from a single analysis that defines an artifact's composition. Nor does characterization consist of sending the artifact for analysis, whatever the type, and recording the results in the object's registration card. Characterization should start, above all, with a careful examination of the object while asking basic questions: What is the information I need? What do I want to know about this object? The answers will help determine the actions required to conserve the object.

From a material viewpoint, characterization aims to provide information about two aspects of the artifact that are essential to its conservation: (1) the constitutive materials and elaboration: materials and techniques used to elaborate the artifact, and (2) the conservation condition: mechanical and/or chemical problems. This is mostly accomplished through a detailed and careful visual inspection followed by an appropriate investigation for confirmation and documentation.

VISUAL INSPECTION

Visual examination will provide an overview of the artifact and stimulate questions and hypotheses about its elaboration and conservation condition (Anderson 1983). It will also help reveal the presence of different materials and point out the key areas to be further analyzed. It is a good idea to have as much knowledge as possible about the object under examination, as well as to be familiar with the types of defects that might normally be encountered (Campbell 2013c).

Constitutive Metals and Elaboration

A thorough examination of the surface aspect of the object together with the magnet test provides an initial indication of the chemical nature of the alloy (table 3.1). However, it is possible that the surface could be a coating that hides another material underneath. For instance, many ferrous alloys are normally coated, because of their low corrosion resistance, and a magnet test would easily reveal the ferrous nature of the alloys. It is therefore important to look carefully for features that are characteristic of coatings, such as blistering, peeling, and defects that expose the base metal (see chaps. 10 and 11).

With regard to elaboration techniques, there is often more than one method employed to shape metals into products. Basically, artifacts are made by casting or wrought processes, usually followed by machining (e.g., turning, boring), joining (e.g., welding, brazing), and finishing processes (e.g., polishing, plating).

Casting can produce a wide variety of products, and the process is capable of producing complex shapes in one piece, incorporating internal cavities or hollow sections. Casting processes are divided into two main categories: expendable, where the mold is sacrificed to remove the part (e.g., sand, investment), and permanent, where the mold is made of metal or refractory ceramic and can be reused (e.g., die, centrifugal). They can be further broken down by the mold material, such as sand or metal, and the pouring method, such as gravity, vacuum, or low pressure. Table 3.2 shows some typical casting processes.

Wrought processes generally change the shape of an existing solid body, initially a plate, sheet, bar, rod, wire, or tubing, which is usually called a workpiece (fig. 3.1). Forming processes that produce wrought artifacts require considerable energy and strong and resilient tools in order to make solids flow (i.e., to achieve plastic deformation). They are often economically viable only for large production volumes.

Table 3.1. | Some characteristics of metallic alloys.

Alloys	Aspect / color		Magnetic	Other general characteristics
	Bare metal	Corrosion products		
Aluminum	Silvery/gray	White, gray Green/blue (Al-Cu)	No	Receives all types of surface finishing (mechanical, chemical, electrochemical, etc.) Lightweight (low density)
Copper	Red, yellow, white	Blue, green, white, brown	No	Receives all types of surface finishing, especially "patina"
Stainless steel	Silvery/gray	Very few products Localized spots of corrosion	Depends on alloy and metallurgical state	Mostly used bare; can be colored by anodization Heavy in comparison with Al alloys
Weathering steel	As-delivered black (from hot mill)	Brown-orange products over the entire surface Exfoliation in wet areas	Yes	Mostly used uncoated (in contrast to carbon steel)
Zinc	Gray, usually dull	White / gray	No	Can be painted or metallic-coated Cast alloys not expensive Fragile fracture

Source: Data from Child 1988, 7–13; National Park Service 2002.

Table 3.2. | Typical casting processes and their applications.

Process	Advantages	Limitations
Sand	Almost any metal can be cast; no limit to part size, shape or weight; low tooling cost	Wide tolerances; relatively coarse surface; some finishing required
Plaster	Intricate part shapes; good dimensional accuracy and surface finish; low porosity	Only nonferrous metals; limited part size and volume of production; mold making time relatively long
Investment (lost-wax)	Almost any metal can be cast; intricate part shapes; excellent surface finish	Part size limited; expensive patterns, molds, and labor
Die	Excellent dimensional accuracy and surface finish; high production rate	High die cost; limited part size; usually limited to nonferrous metals; long lead time

Source: Data from Kalpakjian and Schmid 2014, 257.

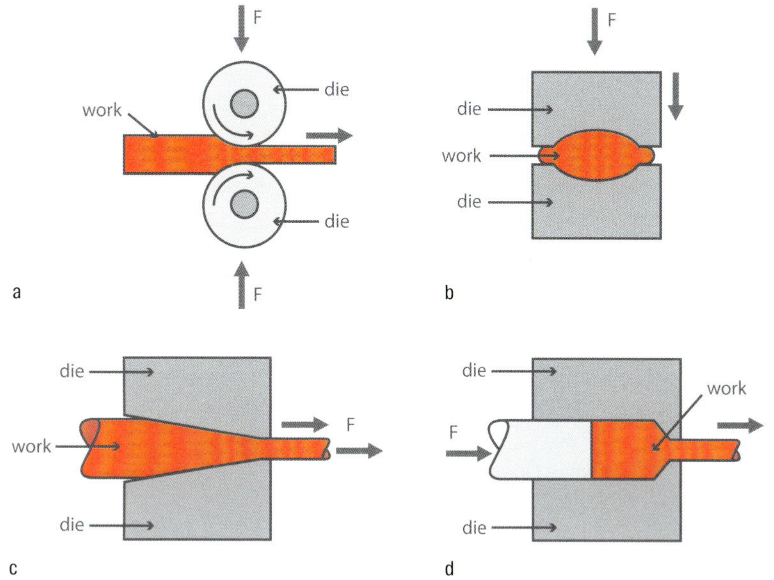

Figure 3.1 | Various wrought processes: Rolling (a); forging (b); extrusion (c); and drawing (d) (F = force) (adapted from Beddoes and Bibby 2003, fig. 3.1).

Moreover, forming may be preferable to casting for other reasons. One reason is the difficulty of manipulating some alloys as liquids, because they have a high melting point (steel) or react with the atmosphere (titanium and magnesium). Another factor is the geometry of the part: products with one dimension significantly different in size from the others (e.g., very long or very thin products) are difficult to cast into the mold and are therefore usually made by a forming process. Finally, the microstructures produced by forming are easier to control and inherently stronger than those produced by casting (dendrites, porosities). This has a direct bearing on the properties of the final product.

Conservation Condition

With regard to conservation condition, modern alloys have to be considered in their particularities, which differ from those of the most traditional alloys. For instance, the precise addition of a number of alloying elements can lead to particular structural features, such as precipitates at grain boundaries, and may result in special forms of corrosion (intergranular, selective). In a number of cases, the surface condition is a fundamental aspect, because alloys such as stainless steel and aluminum are expected to be in a passive condition but may develop localized corrosion, pitting, or crevices if exposed to an aggressive environment.

Modern metallic alloys are very different from what are considered archaeological metals. With rare exceptions—such as the case of an airplane that crashed at sea and remained underwater for many years—most modern metals are not mineralized—that is, they don't have a structure that allows certain types of classical "diseases" or reactivated corrosion.

Knowledge of the materials is desirable, and accurate documentation is especially important in the investigation of corrosion problems. The region of failure itself should be examined visually using a magnifying glass or other suitable viewing equipment. At this stage, one must carefully consider important features, including the extent of damage and the general appearance of the damage zone. If corrosion is suspected, it is important to assess whether it is localized or generalized, uniform or selective, or restricted to an area in contact with other materials. It is critical to record the presence of cracks and note whether they are straight or irregular. The depth of pits and the length of cracks must be measured.

Examination by stereomicroscope can detect small features on corroded surfaces and reveal the manner in which the metal was attacked by the corrosive agent, as well as the degree of corrosion. This information comprises the type of corrosion and whether corrosion was the sole phenomenon involved—or if other mechanisms, such as wear and fracture, were also operative. In addition, in attempting to identify metal alloys from their corrosion products, it is vital to recognize the source of the products: the base alloy, the coating, a joining metal, or possibly external contamination.

ANALYSIS

After careful visual inspection, it is generally possible to postulate a first hypothesis about the materials that compose the artifact. Analytical tools must then be employed to confirm this hypothesis, always keeping in mind that the metallurgical state, not just the composition, is the key information. Moreover, especially in the case of modern alloys, microstructure plays a decisive role in the characterization of the material. Most of these alloys have been developed by adding minute amounts of alloying elements, which induce new microstructural features and remarkably improve the properties but are not detectable by most nondestructive analytical techniques. As discussed in chapter 1, the arrangement of phases and grains bears information about the processing of the part (manufacturing) and the origin of alterations. This knowledge is important for defining adequate conservation conditions.

Constitutive Materials and Elaboration

There are numerous techniques available for the analysis of metallic materials, and, very often, different analytical methods must be used in combination to adequately define the composition and structure of a given artifact. The principles, applications, and limitations of each method are fully described in the specialized literature and are not presented here (see ASM 1986). Some previous knowledge of the expected composition is necessary in order to perform the analysis properly. Most of the analytical methods in use today are instrumental methods whose accuracy depends on calibrations based on alloys of similar composition. Thus, a general idea of the expected elements is required in order to determine the composition.

Chemical spot tests are a valuable aid to alloy identification, if the choice of alternatives is small (ASM 1986). While relatively simple and inexpensive, spot tests require that a collection of chemicals be available, even though only very small amounts of the chemicals are needed to perform them. Moreover, they require some previous experience: the viewer should be trained in using standards and be able to attribute a given feature or color to the presence of a specific element. There is also the problem that it is not always possible to guarantee a completely unequivocal result because of the tests' interference with one another. Finally, even if spot testing is said to be nondestructive, it etches the surface locally where it is performed, leaving marks.

X-ray fluorescence (XRF) is increasingly used in museums and in cultural heritage studies in general because it has the advantage of being nondestructive and portable. This technique is useful for confirming the presence of elements deduced from the visual inspection, but results should be carefully considered to avoid incorrect conclusions. First, if the spatial distribution of the elements in the investigated area is not taken into account, the data may be misinterpreted. For instance, the same average composition may be obtained in two different cases: (1) when detected elements are homogeneously distributed in the bulk, and (2) when the same elements are present in surface layers, as in the case of coatings. This difference is a major consideration, for example, when distinguishing a stainless steel (Fe-C alloy containing Ni and Cr) from a carbon steel with a nickel plus chromium coating. XRF will detect Fe, Ni, and Cr in both cases, although in the first case, the three elements are mixed in the bulk, whereas in the second case, they are distributed in surface layers.

A second point of concern is the elemental percentage as given by XRF or similar analysis, which yields the average value within the region of a measurement. This value should not be considered representative of the whole part. Moreover, such analyses are not quantitative. To perform quantitative analysis, the XRF equipment must be calibrated with a similar specimen—one that not only has a similar amount of the alloying element in question but also has the same matrix element and a similar concentration of the alloying element. Many of the standard or economical methods simply do not detect small amounts, and very high concentrations of elements can result in false readings.

In order to refine the results derived from elemental analysis, in general, microscopic examination of polished and etched specimens is necessary for a complete characterization. Important information may be obtained from one or a few samples as small as a few cubic millimeters taken from carefully selected locations on the artifact.

After polishing and etching, a very complete structural and chemical characterization may be conducted. Metallographic microscopy can reveal the microstructure of the artifact material (Vander Voort 2017; ICCROM 2017), including the grain shape and size, the amount and distribution of phases, and heterogeneities (fig. 3.2).

All of these data contribute to precisely defining the composition of the alloy and its manufacturing sequence. The possible presence of coatings or surface treatments as well as their thickness can also be investigated in detail if sufficient care is taken to preserve the surface during sampling and mounting. The chemical composition of particular phases, inclusions, or other structural features can further be analyzed by energy-dispersive X-ray spectroscopy (EDX), which will give a very precise elemental map of the region (fig. 3.3). Moreover, tests of microhardness, focusing specifically on particular microfeatures, yield complementary insights into mechanical properties (see fig. 2.10).

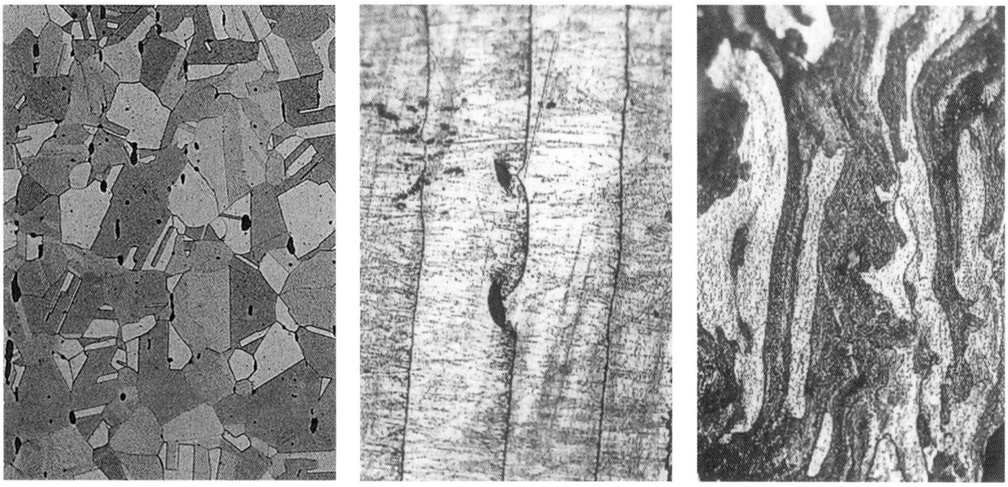

a b c

Figure 3.2 | Microstructure of three copper thin sheets resulting from the fabrication process: Equiaxed and twinned grains indicating deformation and recrystallization of a wrought product (a) (Madeleine's roofing, Paris); four layers, all presenting columnar structure characteristic of electrodeposited copper (b) (Place de la Concorde, Paris); thermal sprayed copper showing irregular alternating layers of metal and darkened oxide, nonbonded interface, and pores (fountain, Cité de Trévise, Paris). Photos: Virginia Costa.

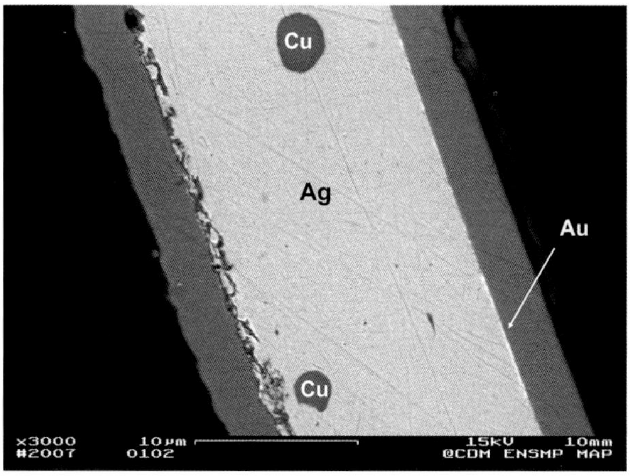

Figure 3.3 | Cross section of a metallic thread sampled from the eighteenth-century *Manteau de chevalier de l'ordre du Saint-Esprit* (Palais Galliera, Paris). Localized EDX analysis allowed determination of the composition, stratigraphy, and distribution of elements. The strip consists of a matrix of pure silver containing dispersed copper globules; the outer surface was gilded with an extremely thin layer of gold (reproduced from Costa, de Reyer, and Betbeder 2012, fig.2).

Conservation Condition

Establishing the nature of a corrosion product on the metal surface is more difficult than analyzing the material, especially when a thorough on-site investigation cannot be made. The surface product can be merely a trace of innocuous soil, a residue of corrodents, or residues of embedded particles. Analysis of corrosion products can be useful only if a specific element is suspected of having contributed to the deterioration.

Microstructural features and conditions such as cracking and crack progression, pit morphology, selective leaching, surface features, and other characteristics can all be investigated by examination of a cross section of a sample. These features provide key evidence regarding the cause of failure and the extent of damage. Corrosion products can be retained on the sample surface by impregnation with a casting-type resin that is allowed to harden before the sample is cut and polished.

Internal discontinuities in parts where damage results from the combined effects of corrosion, stress, and imperfections can be investigated nondestructively by X-radiography and Eddy current technique. Radiography detects changes in density, evidencing voids, inclusions, materials variations, and placement of internal parts (Campbell 2013b). Eddy current detects changes in electrical conductivity caused by material variations, cracks, voids, and inclusions (Campbell 2013a).

A general view of the capabilities of some analytical methods is given in table 3.3.

Table 3.3. | Metals grade identification methods capabilities.

		Methods			
Metal properties		XR spectrometry	Eddy current	Conductivity/ resistivity	Chemical spot test
Chemical composition	Identification/response to specific alloy	E	G	F	E
	Response to surface chemistry	E	F–G	F	E
Mechanical properties	Hardness	N	F	N	N
	Tensile/yield strength	N	F–G	N	N
Monitoring process variables	Cold work	N	F–G	F	N
	Anneal	N	F–G	F	N

Source: Data from ASTM International 2014 (ASTM E1476), 10.
Note: E = excellent, G = good, F = fair, N = not applicable.

REFERENCES AND BIBLIOGRAPHY

Anderson, R. C. 1983. *Inspection of Metals*. Vol. I, *Visual Examination*. Metals Park, OH: American Society for Metals.

ASM [American Society for Metals]. 1986. *Materials Characterization*. ASM Handbook. Metals Park, OH: American Society for Metals.

ASTM International. 2014. *ASTM E1476. Standard Guide for Metals Identification, Grade Verification, and Sorting*. West Conshohocken, PA: ASTM International.

Beddoes, J., and M. J. Bibby. 2003. *Principles of Metal Manufacturing Processes*. Oxford: Elsevier Butterworth-Heinemann.

Campbell, F. C. 2013a. "Eddy Current Inspection.'" In *Inspection of Metals: Understanding the Basics*, 215–32. Materials Park, OH: ASM International.

——. 2013b. "Radiographic Inspection." In *Inspection of Metals: Understanding the Basics*, 233–66. Materials Park, OH: ASM International.

——. 2013c. "Visual Inspection." In *Inspection of Metals: Understanding the Basics*, 21–48. Materials Park, OH: ASM International.

Child, R. 1988. "The Identification of Post-Industrial Revolution Metals." In *Modern Metals in Museums*, edited by R. E. Child and J. M. Townsend, 7–14. London: Institute of Archaeology Publications.

Costa, V., D. de Reyer, and M. Betbeder. 2012. "A Note on the Analysis of Metal Threads." *Studies in Conservation* 57 (2): 112–15.

ICCROM [International Centre for the Study of the Preservation and Restoration of Cultural Heritage]. 2017. *The France-Lanord Collection*. *https://www.iccrom.org/France-Lanord/pages.shtml.*

Kalpakjian, S., and S. R. Schmid. 2014. *Manufacturing Engineering and Technology*. 7th ed. Upper Saddle River, NJ: Pearson.

National Park Service. 2002. "Appendix O: Curatorial Care of Metal Objects." In *The Museum Handbook*. Part 1, *Museum Collections*. Washington, DC: National Park Service.

Vander Voort, G. 2017. *Metallography*. *http://www.georgevandervoort.com/metallography/general/archeometallurgy.html.*

Wilson, M. L. 1986. *ASTM STP550. Nondestructive Rapid Identification of Metals and Alloys by Spot Test*. Philadelphia: ASTM International.

PART II | METALLIC ALLOYS

CHAPTER 4 | ALUMINUM ALLOYS

The unique combination of properties provided by aluminum and its alloys makes them one of the most versatile, economical, and attractive metallic groups for a broad range of uses—from soft, highly ductile wrapping foil to the most demanding engineering applications. The low density combined with relatively high strength have made aluminum alloys the standard material for applications such as aircraft, where specific strength (strength-to-weight ratio) is a major design consideration. Because of their corrosion resistance and good ductility, they are also used for a wide variety of applications including beverage cans; building and construction; and marine, rail, truck, and auto transportation.

One of the first architectural applications of aluminum was in 1884, when a cast aluminum pyramid was mounted on the tip of the Washington Monument after an exhibition in the window of Tiffany's flagship store on Fifth Avenue (Binczewski 1995, 24). The first use of aluminum in sculpture was in the statue of Eros cast in 1892 and erected in Piccadilly Circus, London (*Nature* 1947). In the decade following 1900, the rapid development of new alloys extended the use of aluminum to utensils, marine and aeronautical components, surgical equipment, engineering and scientific instruments, wire products, bottle caps, and electrical conductors (King 1985, 43; Gayle and Look 1992; Van Horn 1967b).

From the 1920s, aluminum was widely used for industrial design, household products, and interior features. At the Bauhaus it was incorporated into lighting fixtures and was particularly favored by designers like Donald Deskey (furniture and lamps in Radio City Music Hall, New York, 1932) and Hans Coray (stacking chair, 1938). In sculpture, aluminum was frequently used in its cast form wherever light weight was required. It was often used by the minimalists in the late 1960s and early 1970s, as it was ideally suited to the manufactured perfection required by their large, simple forms. In the 1960s American artists (Donald Judd, Carl Andre, Frank Stella) produced sculptures using aluminum, chosen for its specific material qualities as a finished surface rather than as a substrate for paintwork. The use of anodizing as a coloring method rather than simply as corrosion protection was also explored during that period (Ward 2008; Byko 2000, 9–10; Bosdorf and Schiek 1973).

The facility with which aluminum may be fabricated into any form is one of its most important assets. It can be cast by any standard process and wrought using any available metalworking techniques. The final properties depend on a complex interaction of chemical composition and microstructural features developed during solidification, thermal treatments, and (for wrought products) deformation processing.

METALLURGICAL BACKGROUND

Pure aluminum cannot be used as a structural material because it is too soft and does not have the required mechanical properties. Alloying elements are added to increase strength, hardness, and resistance to wear. Effects on the properties are specific to the different alloying elements and their combinations and are related to the microstructures that they form as a result of solidification, thermomechanical history, heat treatment, and/or cold working.

The affinity of alloying elements with aluminum is a decisive factor in improving the properties of the alloy. Although no alloying element has complete solid solubility in aluminum, some of them are sufficiently soluble to impart important changes to the original properties (Davis 1993, 31). Moreover, for many of them, there is an appreciable decrease of solubility during cooling (fig. 4.1). This is the fundamental characteristic that provides the basis for substantially improving the hardness and strength of aluminum alloys by

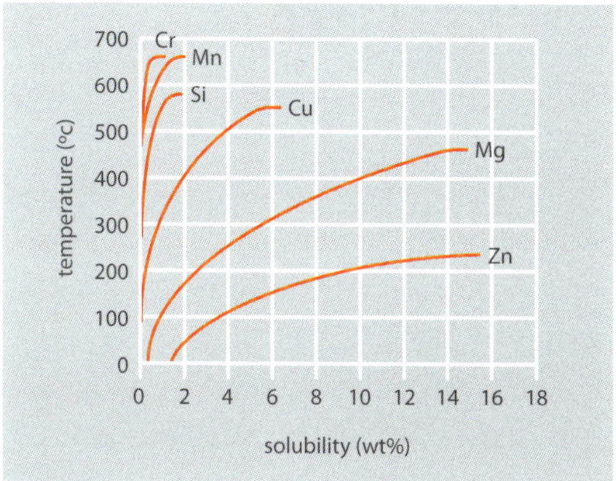

Figure 4.1 | Solid solubility of various alloying elements in aluminum as a function of temperature (adapted from Bray 1990, fig. 5).

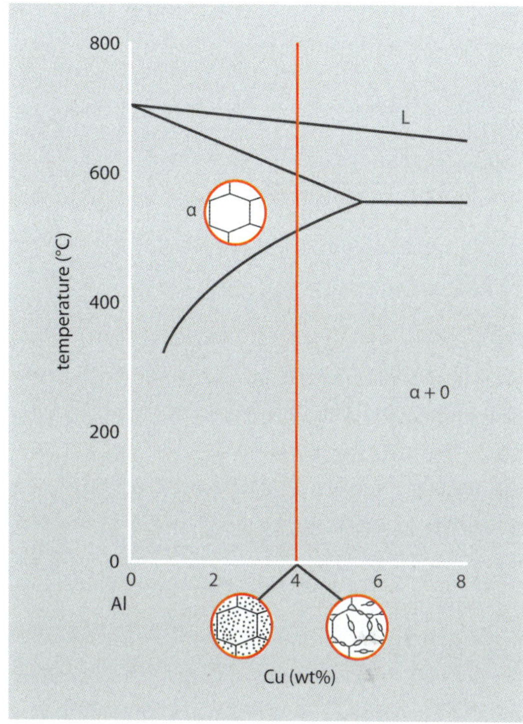

Figure 4.2 | Microstructure of an Al-4%Cu alloy after: solution heat treatment at ~550°C (upper left); moderately fast cooling, producing small and densely distributed precipitates (lower left); and slow cooling, showing large and loosely dispersed precipitates (lower right) (adapted from Ashby and Jones 2006, figs. 10.3, 10.4).

solution heat treatment and subsequent precipitation aging operations (see chap. 2).

Thus, depending on the solubility and consequent possibility of strengthening the material by heat treatment, two groups of alloys can be established (Brown and Willey 1967). The non-heat-treatable group (NHT) is formed by alloys whose mechanical properties cannot be changed by heat treatment, either because added elements are present in very small amounts (e.g., "commercially pure aluminum," Al > 99%) or because they have such a low solubility (Mn, Si) that no remarkable changes in properties can be obtained by varying their concentration. In these cases, strength improvement can be done by solid solution strengthening or work hardening, and it results in a decrease of ductility. These alloys usually present good corrosion resistance.

In contrast, in the heat-treatable group (HT), alloying elements (Cu, Mg, Zn) are rather soluble in the aluminum matrix. They can be dissolved by heating and then, by adequate cooling rates, they can be kept in the crystal matrix at lower temperature, where they would not normally be soluble. They thus form a second phase, improving strength without ductility loss. However, the corrosion resistance of these alloys decreases significantly, because the presence of several phases leads to local potential differences and selective corrosion. Cu is one of the most important additions to aluminum: it has appreciable solubility and a substantial strengthening effect through precipitation hardening (PH) (fig. 4.2).

In conclusion, not only the composition but also the complete thermal and mechanical history significantly influences the properties and characteristics of aluminum-alloy products. For the same chemical composition, depending on the way thermal and mechanical processes have been carried out, the resulting microstructure and, consequently, the mechanical (and corrosion resistance) properties can be completely different. Processes that control strength and other characteristics of both wrought and cast aluminum alloys are called tempering (Kaufman 2000a). They usually consist of a series of treatments (annealing, strain hardening, solution heat treatment) applied in combination in order to obtain specific microstructures and related properties. Table 4.1 presents differences in the properties of the same Al alloy for different metallurgical states as a result of changing the microstructure.

Regarding the surface characteristics, aluminum alloys are highly reactive and combine rapidly with oxygen to form a

Table 4.1. | Typical properties of an Al-Cu alloy (UNS912024) in two different metallurgical states.

Metallurgical state	Yield strength σ_Y (MPa)	Tensile strength σ_T (MPa)	Hardness (HB)
Annealed	95	185	45
Cold-worked and natural-aged	345	485	120

Source: Data from Van Horn 1967a.

protective oxide layer. Therefore, a surface treatment is rarely necessary to provide corrosion protection. However, a great variety of finishes is usually applied to aluminum alloys (see chap. 11, "Anodizing") in order to modify their appearance (color, texture) or enhance other surface characteristics (wear resistance and resistivity) (Flandro and Thomas-Haney 2015).

DESIGNATION, PROPERTIES, AND APPLICATIONS

When dealing with aluminum products, it is very useful to have knowledge of the alloy and temper designation systems, as well as the type of surface finish applied. This information enables an understanding of the manner in which the product was fabricated.

Regarding manufacture, aluminum alloys are usually divided into two major categories: wrought compositions and casting compositions. The designation wrought indicates that the alloys are available primarily in the form of worked products such as sheets, extrusions, tubes, wires, and forgings. Working operations and thermal treatments transform the cast ingot structure into a completely recrystallized or a severely deformed microstructure, which influences properties. Castings can be produced by different processes: die casting and sand casting (see table 3.2) are the most important processes in terms of the volume of production. Smaller quantities of castings are produced in plaster, investment, and composite material molds.

There are alloy compositions and tempers more favorable for wrought compositions or for cast compositions, but a wide range of fabrication possibilities are available for each category. Corresponding alloy nomenclatures have been developed, and their designations change depending on the system adopted (Kaufman 2000e). The wrought alloy designation system of the Aluminum Association (AA) consists normally of four numerical digits: the first digit defines the major alloying class of the series starting with that number; the second digit defines variations in the original basic alloy; and the third and fourth digits designate the specific alloy within the series (AA 2015, 1–29). Moreover the Unified Numbering System for Metals and Alloys (UNS) provides a means of correlating many internationally used metal and alloy numbering systems administered by societies, trade associations, and individual users and producers of metals and alloys. Jointly developed by SAE International and the American Society for Testing and Materials (ASTM), it provides the uniformity necessary for efficient indexing, record keeping, data storage and retrieval, and cross-referencing. The UNS designation number uses the AA designation preceded by "A9" (MatWeb 2017). Table 4.2 shows the meaning of the first digits for both designation systems.

When both high strength and corrosion resistance are required, a composite wrought product—with the trade name Alclad—is proposed (Kaufman 2000c, 188). It is made of metallurgically bonded metal laminates having either commercially pure Al or an aluminum alloy coating that electrochemically protects a high-strength aluminum alloy core (fig. 4.3).

In comparison with wrought alloys, cast alloys contain larger proportions of alloying elements such as silicon and copper, which results in a more heterogeneous structure. Based upon the effects of the primary alloying elements, the alloys with the best castability properties are Al-Si and Al-Si-

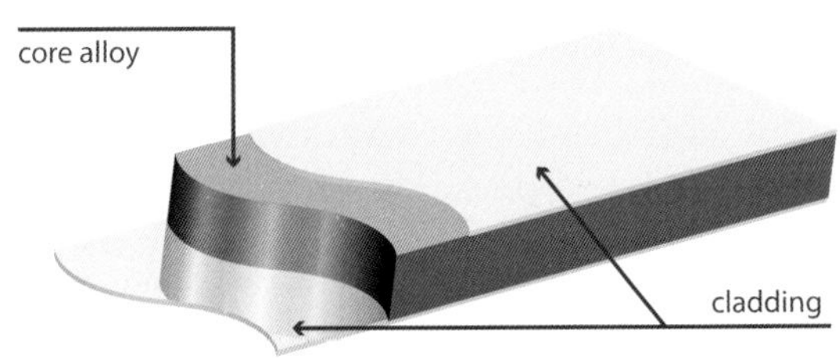

Figure 4.3 | Schematic representation of Alclad.

Table 4.2. | Wrought aluminum-alloy systems: designations, properties, and applications.

Alloy system	UNS (AA)	Properties	Applications
Al (NHT)	A91xxxx (1xxx)	Corrosion resistance; thermal and electrical conductivities; workability	Packaging, chemical equipment, electrical applications
Al-Cu (HT)	A92xxxx (2xxx)	High strength and low corrosion resistance; often clad or painted; usually joined mechanically	Aircraft and truck structures
Al-Mn (NHT)	A93xxxx (3xxx)	High formability and corrosion resistance	Roofing, beverage cans, cooking and chemical equipment
Al-Si (NHT) Al-Si-Mg (HT)	A94xxxx (4xxxx)	Low melting point without brittleness; good flow characteristics	Forgings and welding filler alloys
Al-Mg (NHT)	A95xxxx (5xxx)	Corrosion resistance in marine atmospheres; machinability and weldability	Building, highway structures, storage tanks
Al-Mg-Si (HT)	A96xxxx (6xxx)	Corrosion resistance; extrudability, weldability, and machinability	Architectural and automotive members
Al-Zn (HT)	A97xxxx (7xxx)	High strength and low corrosion resistance; often coated or clad; mechanically joined	Aircraft industry
Other	(8xxx)	Fe, Ni: strength with little loss in electrical conductivity Li: strength and stiffness with low weight	

Source: Data from Kaufman 2000b.
Note: NHT = non-heat-treatable, HT = heat-treatable.

Cu, but these are the worst with regard to finishing, because of the many heterogeneities (Kaufman 2000c, 34; Runge and Chesterfield 2010). The cast alloys designation system also has four digits, and the first digit specifies the major alloying constituent(s), as shown in table 4.3. However, a decimal point is used between the third and fourth digit to make clear that these are alloys in the form of castings. The second and third digits identify the specific aluminum alloy, and the

Table 4.3. | Cast aluminum-alloy systems: designation, properties, and applications.

System	Designation: UNS and (AA)	Properties	Applications
Al (NHT)	A01xxxx (1xx.x)		
Al-Cu (HT)	A02xxxx (2xx.x)	High strength; some high toughness; lack of good surface characteristics	Aerospace industry
Al-Si+Cu or Mg (HT)	A03xxxx (3xx.x)	Fluidity; high strength; some high toughness; readily welded	Aircraft and automobiles (widely used)
Al-Si (HT)	A04xxxx (4xx.x)	Excellent fluidity; good for intricate castings	Computer and dental equipment
Al-Mg (NHT)	A05xxxx (5xx.x)	Corrosion resistance; machinability and good finishing characteristics	Door and window fittings
Al-Zn (HT)	A07xxxx 7xx.x	Machinability; appearance	Furniture, garden tools
Al-Sn (HT)	8xx.x	Machinability (Sn imparts lubricity)	Bearings and bushings

Source: Data from Kaufman 2000b.
Note: NHT = non-heat-treatable, HT = heat-treatable.

fourth digit indicates the product form (0 = castings, 1 = ingot of similar alloy, etc.). For cast alloys, the UNS number is the AA designation preceded by "A0."

The above-mentioned categories of alloys do not provide information about thermal and mechanical processes employed during fabrication, which control the strength and other characteristics of both wrought and cast aluminum alloys. Knowledge about the manner in which the product has been fabricated is provided by the temper designation system. It consists generally of a capital letter indicating the major class of fabrication treatments used, plus one or more numbers providing more specific information about how the processing was carried out (table 4.4). The temper designation is always presented immediately after the alloy designation, separated by a hyphen (e.g., "A92014-T6"). These designations are not intended to provide the exact practices (times, temperatures, reductions) but, rather, the general combinations of usual practices followed. Useful reference material to aid further understanding of aluminum temper designations and the typical mechanical properties of representative wrought and cast aluminum alloys can be found in the specialized literature (Kaufman 2000d).

A third designation system concerns the type of finish applied to the surface of aluminum alloys to enhance appearance or improve functional properties, such as corrosion resistance and adhesion of coatings. The AA designation system for finishes defines seven categories: mechanical finishes, chemical finishes, and five other types of coatings (AA 2003, 1–12). Each of these categories is assigned a letter, and the various finishes in each category are designated by two digits. Anodic coatings can include a third digit to further define the minimum coating thickness and weight (NAAMM and NOMMA 2006; ASM 1994, 571–610). The more common finish subclass designations are presented in table 4.5.

Table 4.4. | Basic temper designations and their applications.

Temper designation	Applications
F (as fabricated)	Applies to products made by shaping processes in which there is no special control over thermal conditions or strain hardening processes employed to achieve specific properties
0 (annealed)	Applies to wrought products that are annealed to obtain the lower-strength temper, usually to increase subsequent workability; applies to cast products that are annealed to improve ductility and dimensional stability
H (strain hardened)	Applies to wrought products that have their strength increased by strain hardening; they may or may not have supplementary thermal treatments to reduce strength and improve ductility Example: H2 = cold-worked and partially annealed
T (thermally treated)	Applies to products that are thermally treated to produce stable tempers other than F or 0, with or without supplementary strain hardening Example: T3 = solution, cold working, and natural aging
W (solution HT)	Applies only to alloys that age spontaneously after solution heat treatment

Source: Data from Kaufman 2000d, 57–59.
Note: HT = heat-treatable.

Table 4.5. | Aluminum alloy finish designations.

Finish designation	Description
M (mechanical)	Removing surface irregularities and smoothing the surface (action is usually abrasive, as in grinding, buffing, blasting, or polishing, in which metal is removed). Applying a texture to the product without material removal (embossing, hammering, and shot peening). Example: M1 = as fabricated (unspecified, specular).
C (chemical)	Accomplished by means of chemicals and generally used for different purposes, such as: merely cleaning the surface without affecting it; providing a clean, matte textured surface (by etching); producing a smooth, bright finish; or chemically converting the surface of the metal, providing a film that is a good substrate for organic coatings. Most of these are intermediate steps in an overall finishing process, requiring at least the subsequent application of some form of protective coating. Example: C4 = chemical coatings (chromate, nonchromate).
A (anodic coatings)	Chemical process in which the surface of the metal is converted to a coating with desirable protective, decorative, or other properties. Properties such as appearance, continuity, abrasion resistance, and porosity are markedly affected both by the purity of the aluminum and by the type and quantity of alloy constituents. Example: A2 = protective and decorative (<10μm or 0.4 mil).
E (electroplated)	Applied to aluminum alloys in order to obtain a specific metallic appearance; to increase resistance to wear, abrasion, or erosion; to increase electrical conductivity; or to improve solderability.
R (resinous, organic)	Providing decorative or protective effects or both, depending on the application.
V (vitreous)	Thin glass films fused onto the metal surface forming a continuous film firmly bonded to the substrate; they are more resistant to abrasion than paint films and nearly as resistant as anodic aluminum oxide finishes.
L (laminated)	Laminated coatings, including veneers, plastic coatings, and bonded films.

Source: Data from AA 2003.

DETERIORATION

Despite the fact that aluminum is a thermodynamically reactive metal, it owes its excellent corrosion resistance to the barrier oxide film that naturally forms on its surface in most environments. For unalloyed aluminum the natural film forms rapidly in air. In the case of aluminum alloys, however, surface reactivity strongly depends on their microconstituents (composition, amount, distribution, and potential relative to the aluminum solid solutions). The effects of composition and microstructure on corrosion are detailed elsewhere for different alloy designations (Davis 1999, 33–35).

As mentioned above, the metallurgical and thermal treatments used to tune the mechanical properties of aluminum alloys also influence their corrosion resistance. This is because they locally influence the distribution of the constituents and the magnitude of residual stresses. If one portion of an alloy surface has a structure different from the remainder, its potential will be different, leading to preferential corrosion of the less noble part. Even the heat of welding can produce such inhomogeneities.

Surface condition also plays an important role in corrosion resistance. Scratches or abrasions disrupt the naturally formed or produced protective surface layer, creating local attack sites.

Corrosion of aluminum alloys, like that of other metals, can be of many types, and its identification helps in determining its cause and in recommending or developing methods of control. Measurement of the extent of attack and comparisons among metals and alloys must be based on recognition of the predominant type of attack (Reinhart 1969).

At room temperature, aluminum is protected by its oxide, which is stable in the pH range of about 4.0–8.5. This passive state, however, varies somewhat with temperature and the presence of chemical substances that can form soluble complexes or insoluble salts with aluminum. In particular, in certain strongly acidic or strongly alkaline electrolytes, the oxide film is dissolved, and uniform attack of the metal occurs (fig. 4.4). Uniform attack can be recognized early and remedied relatively easily.

Localized corrosion, on the other hand, occurs where the passive layer is locally broken, usually because of poor or inhomogeneous access of oxygen and the presence of halide

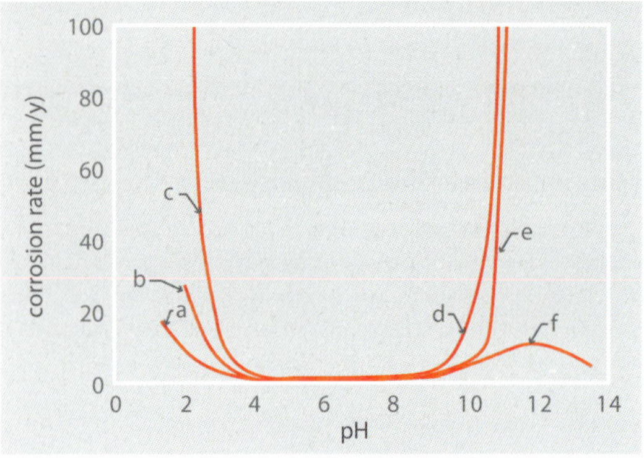

Figure 4.4 | Effect of pH on corrosion of 1100-H14 alloy by various chemical solutions: Hydrochloric acid (a); phosphoric acid (b); hydrofluoric acid (c); sodium carbonate (d); sodium hydroxide (e); and ammonium hydroxide (f). Outside the pH range of 4.5–8.5, where the corrosion rate is minimal, the nature of the individual ions in solution is more important than the degree of acidity or alkalinity (adapted from Van Horn 1967a, fig. 34).

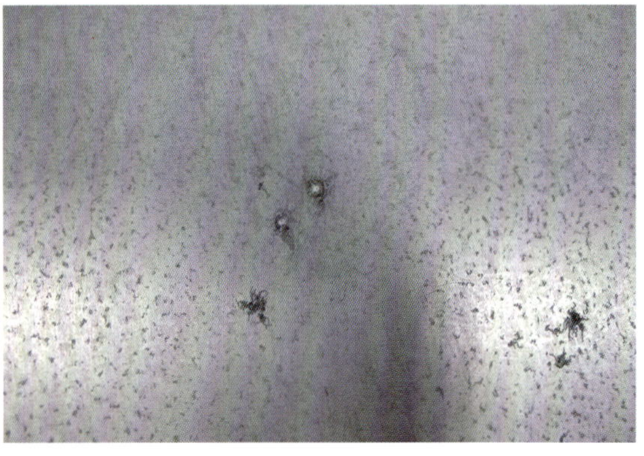

Figure 4.5 | Pitting on aluminum alloy from the car prototype Le Carrosse by Paul Arzens, 1951, Musée des Arts et Métiers, Paris (reproduced from Florescu 2015, table 4a).

anions (e.g., chlorides) in the environment (fig. 4.5). In such conditions, regions where aeration is ineffective (crevices or pits) are rich in such ions and prone to enhanced corrosion because their potential is anodic with respect to other regions (see chap. 2, "Forms of Corrosion") (Davis 1999, 32).

Pitting occurs at discontinuities in the oxide film on aluminum exposed to atmosphere, freshwater or saltwater, or other neutral electrolytes. The size, geometry, and distribution of pits depend upon alloy composition, oxide film quality, and the nature of the corroding agents. The resistance of aluminum to pitting depends significantly on its composition: pure Al is more resistant than any of the alloys Al-Zn7xxx and Al-Cu 2xxx. Among the alloys, those containing higher copper content are the least resistant to pitting.

A local cell is also created when an electrolyte is present in a crevice formed between two faying surfaces (surfaces fastened at a joint)—the aluminum and another material, metallic or not. Depending on the nature of the material in close contact with the aluminum surface and on environmental conditions, corrosion can take different forms, such as water staining and poultice or filiform corrosion.

Water staining occurs when water is present in the restricted space between two layers of aluminum in close proximity, as during storage or transit, because of inadequate prevention of rain ingress or humidity condensation. Grayish patches stain the shiny surface and impair the use of the sheet where surface appearance is important, although in such cases, corrosion is shallow, and perforation rarely occurs.

Poultice corrosion occurs beneath hygroscopic attachment or insert (lamination of paper, cloth, or wood) and under deposits of road debris and hygroscopic particulate matter. These areas tend to remain wet, resulting in a highly corrosive environment owing to the moisture entrapment effect of the poultice. The products formed during weathering of unprotected aluminum alloys in most natural environments provide a base for the accumulation of airborne particles of dirt and soil, and the general aspect of the surface will vary from a pleasant gray appearance (relatively dirt-free environments) to a dark or even black surface finish as a result of soil accumulation (industrial areas, horizontal surfaces). Moreover, even if innocuous in itself, the accumulation of dirt on the surface may lead to corrosion, either by holding humidity for longer times in these sites or by forming deposits that will promote differential aeration and induce localized corrosion. A study conducted to evaluate the conservation condition of an outdoor art panel made of anodized aluminum showed that regions near the top, sheltered from rain by a kind of eaves, presented the highest pitting density (approx. 30%–50% of the surface), while the regions of the lowest panels, exposed to weather, exhibited a very low pit density (approx. 5%). Figure 4.6a shows a detail of a heavily corroded area.

Filiform corrosion occurs on an aluminum surface under a thin organic coating, usually when exposed to warm and humid atmospheres. This form of corrosion generally initiates at defects or mechanical scratches in the coating and grows along the metal-coating interface as threadlike filaments. The

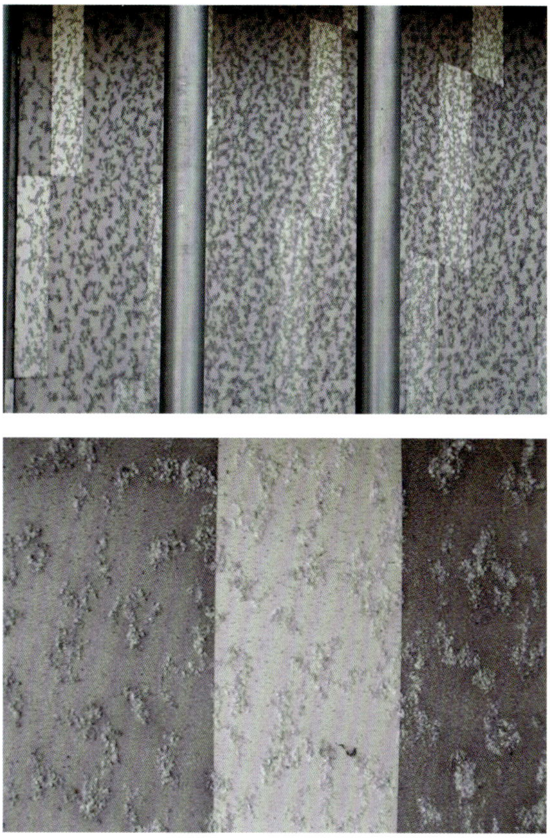

a

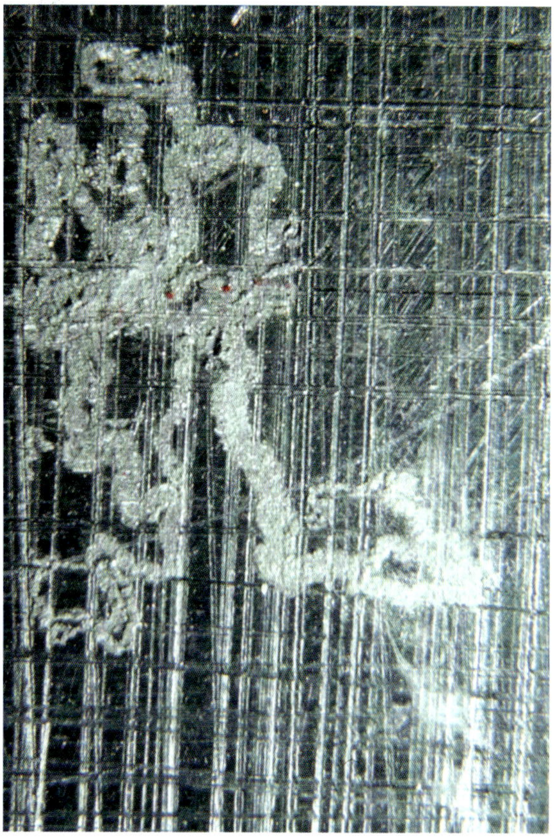

b

Figure 4.6 | Poultice corrosion: Details of a sheltered region of an aluminum outdoor panel by Yvaral (2003), Fontenay-sous-bois (a); filiform corrosion on aluminum sheet from *The Frame* by Frida Kahlo (1937), Musée National d'Art Moderne, Paris (b). Photo: Virginia Costa (a); reproduced from Neuner 2003 (b).

corrosion products cause a bulge in the surface coatings. The surface film itself is not involved in the process, except insofar as it contains zones of poor adhesion that form the crevices in which corrosion occurs upon exposure to moisture. The presence of oxygen is fundamental because it supplies the primary reactant for the cathodic reaction—the anodic area being the head of the filament. Considerable acidity is generated at the leading edge of the head, which has also been reported to contain a large concentration of chloride (fig. 4.6b).

Selective corrosion occurs when particular areas of the microstructure, such as the grain boundaries or closely adjacent regions, are selectively attacked without appreciable corrosion of the grains themselves. The mechanism differs with various alloys systems, the structure, and the thermal treatment.

Intergranular corrosion commonly occurs in alloys where the solute elements of the alloy are not kept in solid solution. Instead, they precipitate at grain boundaries, developing a constituent-depleted region, which creates a potential difference between the zones. This leads to preferential corrosion of the less noble areas when the surface is exposed to a harmful environment. Depending on the alloy family, the precipitates can be anodic (Al-Mg, Al-Zn-Mg) or cathodic (Al-Cu) to the matrix (depleted solid solution). In either case, selective attack of the grain boundary region occurs. The degree of intergranular corrosion depends on the quantity, size, and distribution of second-phase intermetallic precipitates, which are a function of fabrication practices. Alloys that do not form second-phase microconstituents at grain boundaries or those in which the precipitates have corrosion potentials similar to the matrix are not susceptible to this type of corrosion.

Identifying intergranular corrosion requires the examination of a cross section with a microscope, because corrosion is limited to the immediate grain boundary region.

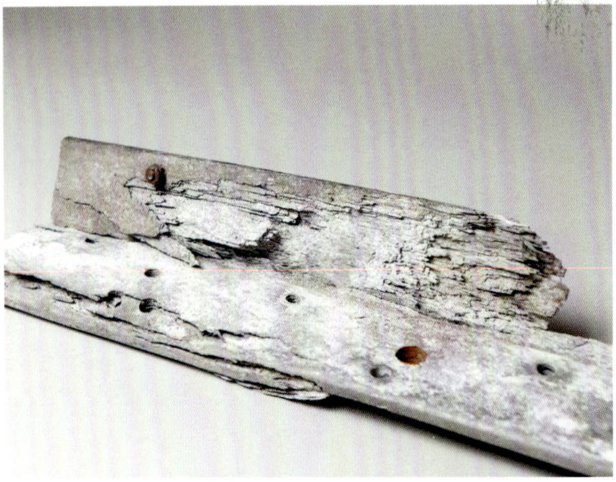

Figure 4.7 | Extruded aluminum alloy from the Dymaxion House by Buckminster Fuller, showing extensive exfoliation. Photo: © J. Paul Getty Trust.

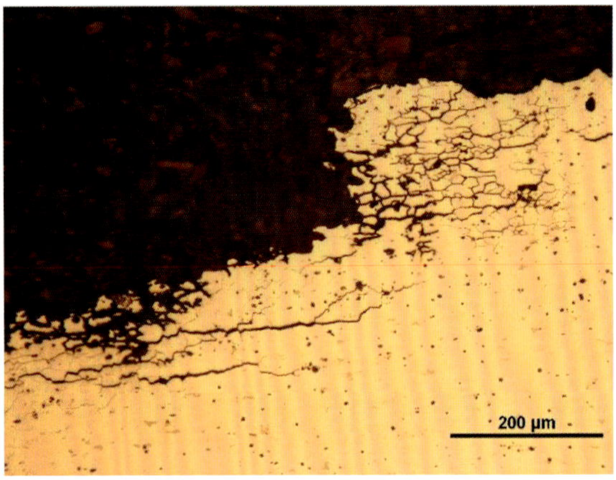

Figure 4.8 | Intergranular path of stress-corrosion cracks on an Al-Cu sheet from an airplane after many decades in the sea (M. Brunet, CEMES-CNRS, unpublished data, 2016).

It penetrates quickly but reaches a self-limiting depth (owing to limited transport of oxygen), spreading laterally over the entire surface—in contrast to pits, which are often confined to discrete sites. Evaluation by weight loss or visual observation is unreliable.

Another type of selective corrosion found in aluminum alloys is exfoliation. It is also referred to as layer, stratified, or lamellar corrosion and results from a relatively rapid lateral attack along electrochemically anodic strata parallel to the metal surface. Layers of uncorroded metal between the selective paths are split apart by the voluminous corrosion products formed along the paths of attack (fig. 4.7). Exfoliation occurs predominantly in products that have markedly directional structures in which highly elongated grains form platelets that are thin relative to their length. Potential differences between the strata can result from the presence of aligned intergranular precipitates or strata that differ slightly in composition. In certain products, such as extrusions and die forgings of high-strength aluminum alloys, the large recrystallized grains forming the "skin" (surface layer) on these products are usually slightly cathodic to the underlying fragmented (unrecrystallized) grain structure. Exfoliation is a very deleterious form of corrosion because the splitting action continuously exposes film-free metal and the corrosion is not self-limiting, which rapidly reduces the load-carrying ability of the structure. Susceptible alloys are the heat-treatable Al-Cu and Al-Zn-Mg-Cu, where exfoliation is confined to relatively thin sections of highly worked products with an elongated grain structure.

Stress corrosion cracking (SCC) is a combination of corrosion and cracking in or adjacent to grain boundaries, oriented approximately normal both to the surface and to the direction of a sustained tension stress. In aluminum alloys, this form of corrosion is usually intergranular and is influenced by many factors, including metal composition and structure, magnitude of stress, and environment. Since SCC can be a consequence of precipitation in grain boundaries, it is greatly influenced by metallurgical treatment. Evidence for SCC can be obtained by metallographic examination, which will reveal secondary intergranular cracks in the immediate vicinity of the main failure (fig. 4.8). SCC has been seen in higher-strength alloys of the Al-Cu, Al-Zn-Mg-Cu, Al-Mg (>3% Mg), and Al-Si-Mg (high-silicon) types.

Under many environmental conditions, aluminum and its alloys act as the anode in galvanic cells when in contact with most other metals, protecting them by corroding sacrificially. Contact between aluminum and more noble metals should therefore be avoided, especially in any environment in which aluminum by itself is subject to pitting corrosion.

CHARACTERIZATION

Aluminum and its alloys are nonmagnetic and can be differentiated from other metals by their low density and good resistance to general corrosion. However, differentiating between various classes of aluminum alloys is very difficult, even on the basis of a chemical composition determination, because the composition of alloys may differ by only a few parts per thousand of alloying elements. Moreover, it is hard to assign an alloy to any particular designation, as most of them have alloying elements within a range where various grades overlap.

Fortunately, each alloy system presents particular microstructural features and properties that are characteristic of the group. These are just as important as the chemical composition for understanding an artifact in its context—its origin, history, and uses (Hunter, Montgomery, and Wilcox 1967, 59). One such feature is the presence of characteristic elements, such as constituent particles and precipitates, which are only visible in metallographic sections.

Constituent particles are generally formed between aluminum and an impurity or alloying elements during solidification. These particles are an integral part of the alloy and are not considered inclusions. Some particles can be identified by their shape, hardness, or characteristic color, whereas others can be distinguished by etching characteristics. Precipitate structures are highly significant because they frequently indicate the metallurgical condition, the mechanical characteristics, and the corrosion behavior of the alloy. The precipitates are fine particles of phases that were in solution in aluminum melt at an elevated temperature and precipitated from solid solution at a lower temperature. Precipitates form at grain or fragment boundaries, dislocations, and vacancies. Their size, shape, and location yield information about the thermal conditions that led to their formation.

The phases that appear in aluminum alloys may be the alloying elements themselves (e.g., silicon or lead), compounds that do not necessarily contain aluminum (e.g., Mg_2Si), or compounds that contain aluminum and one or more alloying elements (e.g., $CuAl_2$). Figure 4.9 shows examples. A detailed list providing information on the identification of the most common phases of commercial alloys can be found elsewhere (Stevens, 1985, 356–57).

Aluminum alloy castings have a dendritic structure directly related to solidification rate: the faster the rate, the finer the structure. As solidification rate increases, the size of the solid-solution dendrite branches decreases, and the size of constituent particles becomes finer. Thus, die castings are generally finer in structure than permanent mold castings, which in turn are finer than sand castings. Aluminum castings generally contain some voids—a term that designates a number of casting discontinuities of various origins. The most prevalent types are gas porosity, shrinkage porosity, entrapped gas holes, and cracks (Warmuzek 2004).

In addition, stratification and directionality of structures relate to the type and amount of metal flowing during fabrication, particularly the type and degree of working. Wrought products exhibit recrystallized or unrecrystallized structures, depending on the composition, type, amount,

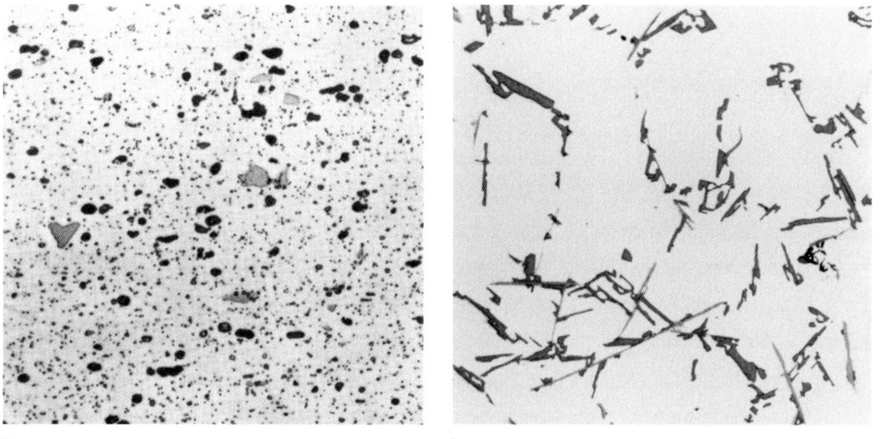

a b

Figure 4.9 | Alloy 2024-O: Light gray particles of $(Cu,Fe,Mn)Al_6$, large black particles of $CuMgAl_2$, and fine particles of $CuMgAl_2$ (Etch 25% HNO_3. 500X) (a); alloy 356-T51: The angular, dark gray constituent is silicon; the black script is Mg_2Si; the blades are $Fe_2Si_2Al_9$; and the light script is $FeMg_3Si6Al8$ (Etch 0.5% HF. 250X) (reproduced from Stevens 1985, figs. 48, 125).

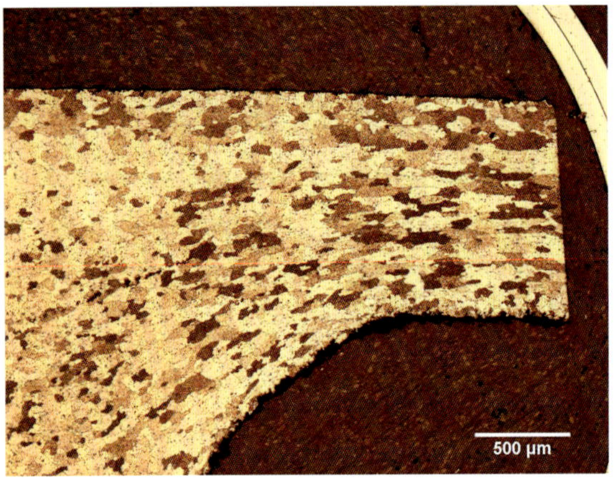

Figure 4.10 | Cross section of an Al-Cu alloy part from an aircraft. The form and distribution of the grains reveals the deformation history of the part (M. Brunet, CEMES-CNRS, unpublished data, 2016).

and temperature of working and on any thermal treatments applied (fig. 4.10).

The surface of aluminum-alloy artifacts—like the presence of fine lines on the surface of extruded parts, die lines for castings, and so on—can also provide information about manufacturing techniques. With regard to the presence of anodic film on the surface, AA recommends some simple tests as guides for determining the type of finish on aluminum: for instance, when a bare surface is rubbed with an eraser, it may lighten in color while the eraser darkens. For an anodized surface, however, no change will be observed (AA 2008, 1–24).

Some characteristic problems that occur during manufacture can be identified with the aid of the visual charts (AA 2002, 1–52). A detailed evaluation of corrosion damage, however, requires analytical methods described by standards, such as metallographic examination to determine the extent of attack underneath the surface (Lifka 2005).

CONSERVATION-RELATED LITERATURE

Publications related to the conservation of artifacts made of aluminum alloys are few and quite diverse. A literature survey identifies some review articles (Kelley 2014; Staveteig 2008; Adams and Hallam 1993; Duncan 1988; Meehan 2002) and a few theses on specific subjects, including the conservation of an aircraft wreck (Gujarathi 2008) and a car prototype (Florescu 2015), as well as filiform corrosion on aluminum substrate for painting (Neuner 2003). Case study publications, sometimes associated with scientific research, are more numerous, covering cameras (Berger 2008), buildings (Trentelman, Ashby, and Donlon 2002; Ashby 2000; Walker 2001; Jandl 1991), metal threads (Ventelä 2013), aircraft wrecks (MacLeod 1983; Goodway and Leyes 1993; Magnin et al. 1997; Degrigny 1995; Rocca and Mirambet 2010), and sculptures (Martini et al. 2012; Larson 1995; Marshall and Sturman 1993; Mcdonald, van der Molen, and Rolfe 2013; Williams and Lowinger 1995; Belman 2002; Belman, Mack, and Sturman 2006; Viseux 2010). However, case study publications often lack important information, such as the characterization of the alloy, the reasons a particular intervention was chosen, and even the results obtained.

A 2014 ICOM-CC conference, Aluminum: History, Technology and Conservation, featured papers contributing to different aspects of aluminum conservation: history, production, fabrication, and use of aluminum; corrosion and deterioration studies; materials characterization and identification; conservation of archaeological objects; conservation and use of aluminum alloys in contemporary art; conservation of aluminum architectural structures and elements; coatings, surface treatments, and corrosion inhibition; and preventive conservation. There is also an ongoing research project concerning the study, identification, and classification of corrosion forms of aluminum artifacts (Degrigny et al. 2016).

REFERENCES AND BIBLIOGRAPHY

AA [Aluminum Association]. 2002. *Visual Quality Characteristics of Aluminum Sheet and Plate. QCA-1.* Washington, DC: Aluminum Association.

——. 2003. *Designation System for Aluminum Finishes. DAF-45.* Arlington, VA: Aluminum Association.

——. 2008. *Care of Aluminum. CA-92.* Washington, DC: Aluminum Association.

——. 2015. *International Alloy Designations and Chemical Composition Limits for Wrought Aluminum and Wrought Aluminum Alloys.* Arlington, VA: Aluminum Association.

Adams, C., and D. Hallam. 1993. "Finishes on Aluminium: A Conservation Perspective." In *Saving the Twentieth Century: The Conservation of Modern Materials*, edited by D. W. Grattan, 273–85. Ottawa: Canadian Conservation Institute.

Ashby, J. 2000. "Re-discovering a Dwelling Machine: Buckminster Fuller's Dymaxion House (1928–45)." *Docomomo Journal*, no. 22:36–39.

Ashby, M. F., and D. R. H. Jones. 2006. *Engineering Materials 2. An Introduction to Microstructures, Processing and Design.* Oxford: Butterworth-Heinemann.

ASM International. 1994. *Cleaning and Finishing of Aluminum and Aluminum Alloys.* Vol. 5, *Surface Engineering.* ASM Handbook. Materials Park, OH: ASM International.

Belman, M. 2002. "Three Case Studies of Outdoor Sculpture with Problematic Intent Issues." *AIC Objects Specialty Group Postprints* 9:53–67.

Belman, M., A. Mack, and S. Sturman. 2006. "The 2004–2005 Treatment of Alexander Calder's Last and Largest Mobile, 'Untitled 1976.'" *AIC Objects Specialty Group Postprints* 13:45–65.

Berger, O. 2008. "Rapport concernant le traitement de conservation-restauration d'une série de petits appareils photographiques pour pigeons, Brevet Michel, 1930." Technical Report. Basel: Art Metal Conservation. *http://www.artmetalconservation.com/pdf/dos_apphoto.pdf.*

Binczewski, G. J. 1995. "The Point of a Monument: A History of the Aluminum Cap of the Washington Monument." *JOM* 47 (11): 20–25.

Bosdorf, L., and H. Schiek. 1973. "Verarbeitung von Aluminium bei der Plastikenherstellung." *Metall* 27 (5): 471–74.

Bray, Jack W. 1990. "Aluminum Mill and Engineered Wrought Products." In *Properties and Selection: Nonferrous Alloys and Special-Purpose Materials*, 29–61. ASM Handbook, vol. 2. Materials Park, OH: ASM International.

Brown, R. H., and L. A. Willey. 1967. "Constitution of Alloys." In *Properties, Physical Metallurgy and Phase Diagrams*, edited by K. R. Van Horn, 31–54. Vol. 1 of *Aluminum.* Metals Park, OH: American Society for Metals.

Byko, M. 2000. "Aluminum Exhibits Its Versatility in Art, Life." *JOM* 52 (11): 9–12.

Davis, J. R., ed. 1993. *Aluminum and Aluminum Alloys.* ASM Specialty Handbook. Materials Park, OH: ASM International.

——. 1999. "Understanding the Corrosion Behavior of Aluminum." In *Corrosion of Aluminum and Aluminum Alloys*, edited by J. R. Davis, 25–43. Materials Park, OH: ASM International.

Degrigny, C. 1995. "Stabilisation de moteurs d'avions immergés." *Studies in Conservation* 40 (1): 10–18.

Degrigny, C., J. Schröter, M. Ledergerber, C.-A. Künzi, O. Sereda, C. Bilat, and F. Hachez-Leroy. 2016. "The EthICAL Project: Study, Identification and Classification of Corrosion Forms of Aluminium Artefacts." In *Metal 2016: Proceedings of the Interim Meeting of the ICOM-CC Metals Working Group, September 26–30, New Delhi, India*, edited by R. Menon, C. Chemello, and A. Pandya, 333. Paris: ICOM-CC.

Duncan, S. 1988. "Aluminium: Its Alloys, Coatings and Corrosion." In *Modern Metals in Museums*, edited by R. E. Child and J. Townsend, 27–32. London: Institute of Archaeology Publications.

Englehart, E. T. 1967. "Cleaning and Maintenance of Surfaces." In *Fabrication and Finishing*, edited by K. R. Van Horn, 772–74. Vol. 3 of *Aluminum.* Metals Park, OH: American Society for Metals.

Flandro, X., and H. M. Thomas-Haney. 2015. "A Survey of Historic Finishes for Architectural Aluminum, 1920–1960." *Journal of Preservation Technology* 46 (1): 13–21.

Florescu, M. 2015. "Voyage en Carrosse: Étude et conservation-restauration d'un prototype de micro-car en aluminium de Paul Arzens, 1951—Évaluation de différents traitements des surfaces en aluminium, en vue de l'obtention d'un rendu uniforme des finitions d'origine." MSc diss., Institut National du Patrimoine.

Gayle, M., and D. W. Look. 1992. "Aluminum." In *Metals in America's Historic Buildings: Uses and Preservation Treatments*, edited by M. Gayle and J. C. Waite, 84–87. Washington, DC: National Park Service.

Goodway, M., and R. A. Leyes. 1993. "Back to Kitty Hawk: Investigating the First Aircraft Aluminum." *JOM* 45 (11): 20–22.

Gujarathi, K. 2008. "Corrosion of Aluminum Alloy 2024 Belonging to the 1930s in Seawater Environment." MSc diss., Texas A&M University.

Hunter, M. S., A. M. Montgomery, and G. W. Wilcox. 1967. "Microstructure of Alloys and Products." In *Properties, Physical Metallurgy and Phase Diagrams*, edited by K. R. Van Horn, 55–78. Vol. 1 of *Aluminum.* Metals Park, OH: American Society for Metals.

Jandl, H. W. 1991. "With Heritage So Shiny: America's First All-Aluminum House." *APT Bulletin* 23 (2): 38–43.

Kaufman, J. G. 2000a. "Aluminum Alloy and Temper Designation Systems of the Aluminum Association." In *Introduction to Aluminum Alloys and Tempers*, edited by J. G. Kaufman, 9–22. Materials Park, OH: ASM International.

——. 2000b. "Applications for Aluminum Alloys and Tempers." In *Introduction to Aluminum Alloys and Tempers*, edited by J. G. Kaufman, 87–114. Materials Park, OH: ASM International.

——. ed. 2000c. *Introduction to Aluminum Alloys and Tempers.* Materials Park, OH: ASM International.

——. 2000d. "Understanding the Aluminum Temper Designation System." In *Introduction to Aluminum Alloys and Tempers*, edited by J. G. Kaufman, 39–76. Materials Park, OH: ASM International.

——. 2000e. "Understanding Wrought and Cast Aluminum Alloys Designation." In *Introduction to Aluminum Alloys and Tempers*, edited by J. G. Kaufman, 23–37. Materials Park, OH: ASM International.

Kelley, S. J. 2014. "Aluminum." In *Twentieth-Century Building Materials: History and Conservation*, edited by T. C. Jester, 13–18. Los Angeles: Getty Conservation Institute.

King, Frank. 1985. "Aluminium 1: Aluminium Alloys, 1900–1950." *Historical Metallurgy* 19 (1): 39–43.

Larson, J. 1995. "Eros: The Laser Cleaning of an Aluminium Sculpture." In *From Marble to Chocolate: The Conservation of Modern Sculpture. Tate Gallery Conference, 18–20 September 1995*, edited by J. Heuman, 53–58. London: Archetype Publications.

Lifka, B. W. 2005. "Aluminum (and Alloys)." In *Corrosion Tests and Standards: Application and Interpretation*, 2nd ed., edited by R. Baboian, 547–57. West Conshohocken, PA: ASM International.

MacLeod, I. 1983. "Stabilization of Corroded Aluminum." *Studies in Conservation* 28 (1): 1–7.

Magnin, M., G. Biousse, O. El Kedim, and P. Fluzin. 1997. "Expertise de l'endommagement, restauration et conservation de l'avion prototype 'Concorde 001.'" In *Metal 95: Proceedings of the International*

Conference on Metals Conservation, Semur-en-Auxois, 25–28 September 1995, edited by I. MacLeod, S. Pennec, and L. Robbiola, 297–304. London: James and James.

Marshall, A., and S. Sturman. 1993. "Conservation of One of Alexander Calder's Largest Mobiles." In *Saving the Twentieth Century: The Degradation and Conservation of Modern Materials*, 301–6. Ottawa: Canadian Conservation Institute.

Martini, C., B. Ballarin, C. Chiavari, and A. Roversi. 2012. "The Aluminum-Cast Madonna Statue of 'Tempio Votivo,' Lido di Venezia (Italy): Identification of Degradation Factors and Assessment of a Cleaning Procedure." *Materials Chemistry and Physics* 137 (1): 404–13.

MatWeb. 2017. "Metal Alloy UNS Number Search." *http://www.matweb.com/search/SearchUNS.aspx*.

Mcdonald, E., R. van der Molen, and M. Rolfe. 2013. "When Modern Metals Meet Modern Museums." In *Metal 2013: Edinburgh, Scotland: Interim Meeting of the ICOM-CC Metals Working Group, 16–20 September 2013, Conference Proceedings*, edited by E. Hyslop, V. Gonzalez, L. Troalen, and L. Wilson, 147–50. Edinburgh: Historic Scotland.

Meehan, P. 2002. "Aluminium and Its Alloys: Current Approaches to Conservation." In *Back to Basics: Selected Papers From a Series of Conferences Organised by the Metals Section of UKIC, 1999–2000*, edited by H. Moody, 13–15. London: Metals Section Press.

NAAMM and NOMMA [National Association of Architectural Metal Manufacturers and National Ornamental and Miscellaneous Metals Association]. 2006. "Aluminum." In *Metal Finishes Manual for Architectural and Metal Products*, 1-1 to 1-22. Chicago: National Association of Architectural Metal Manufacturers and National Ornamental and Miscellaneous Metals Association. *http://www.nomma-northeast.org/wp-content/uploads/2010/11/NAAMM-NOMMA-Finishes-Manual.pdf*.

Nature [journal]. 1947. "Statue of Eros in Piccadilly Circus, London." *Nature* 159:870.

Neuner, M. 2003. "Une conduite dans le travail toute contraire à celle qu'on pratique d'ordinaire. La peinture sous verre: Sa nature, sa technologie et ses problématiques de conservation-restauration. Étude et traitement de 'Saint-Sépulcre,' anonyme, XIXe siècle, Musée des Arts et Traditions Populaires, Paris, et de 'Le Frame,' Frida Kahlo, 1938, Musée National d'Art Moderne, Paris. Recherche technico-scientifique sur les fixatifs à base de cire." MSc diss., Institut National du Patrimoine.

Reinhart, F. M. 1969. *Corrosion of Materials in Hydrospace.* Part 5, *Aluminum Alloys*. Technical Note N-1008. Port Hueneme, CA: Naval Civil Engineering Laboratory.

Rocca, E., and F. Mirambet. 2010. "The Electrochemical Techniques for the Diagnosis and Restoration Treatments of Technical and Industrial Heritage: Three Examples of Metallic Artefacts." *Journal of Solid State Electrochemistry* 14 (3): 415–23.

Runge, J., and L. Chesterfield. 2010. "The Science of Successfully Anodizing Die Cast Substrates." *Aluminium International Today* [*AIT*], Nov.–Dec.: 30–36.

Staveteig, K. R. 2008. *Maintenance and Repair of Historic Aluminum Windows*. Preservation Tech Notes, Windows 22. Washington, DC: National Park Service.

Stevens, R. H. 1985. "Aluminum Alloys." In *Metallography and Microstructures*, 9th ed., edited by ASM Handbook Committee, 351–88. Metals Park, OH: American Society for Metals.

Trentelman, K. A., J. Ashby, and W. T. Donlon. 2002. "The Characterization and Conservation of Aged Aluminum Alloys: Buckminster Fuller's Dymaxion House." In *Materials Issues in Art and Archaeology VI: Symposium Held November 26–30, 2001, Boston, Massachusetts, USA*, edited by P. B. Vandiver, M. Goodway, and J. Mass, 3–14. Warrendale, PA: Materials Research Society.

Van Horn, K. R., ed. 1967a. *Aluminum.* Vol 1, *Properties, Physical Metallurgy and Phase Diagrams*. Metals Park, OH: American Society for Metals.

———. ed. 1967b. *Aluminum.* Vol. 2, *Design and Applications*. Metals Park, OH: American Society for Metals.

Ventelä, A.-S. 2013. "Metalleffektfäden des 20. Jahrhunderts: Identifizierung und Analyse von laminierten Metallfäden." *Beiträge zur Erhaltung von Kunst- und Kulturgut* 2:8–15.

Viseux, A.-C. 2010. "La Sculpture monumentale en aluminium: La restauration du groupe sculpté *Les Nageurs* à la Courneuve (Seine-Saint-Denis)." In *Monumental*, semestriel 2:100–1.

Walker, D. 2001. "The Dymaxion." *Advanced Material and Processes* 159 (11): 44–48.

Ward, G. W. R., ed. 2008. "Aluminum." In *The Grove Encyclopedia of Materials and Techniques*, 8–10. New York: Oxford University Press.

Warmuzek, M. 2004. "Metallographic Techniques for Aluminum and Its Alloys." In *Metallography and Microstructures*, edited by G. F. Vander Voort, 711–75. Materials Park, OH: ASM International.

Williams, D., and R. Lowinger. 1995. "Quiet Collaboration: The Special Relationship between Artists and Their Fabricators." In *From Marble to Chocolate: The Conservation of Modern Sculpture. Tate Gallery Conference, 18–20 September 1995*, edited by J. Heuman, 130–35. London: Archetype Publications.

CHAPTER 5 | COPPER ALLOYS

Copper and copper alloys constitute an important group of commercial metals. They are widely used because of their excellent electrical and thermal conductivities, outstanding resistance to corrosion, ease of fabrication, and good strength and fatigue resistance. Copper alloys are readily cast into cakes, billets, rods, or plates that are suitable for subsequent hot or cold processing via all the standard rolling, drawing, and joining methods. The majority of copper alloys are highly workable hot or cold, making them readily available commercially in various forms: forgings, bars, wires, tubes, sheets, and foils.

As with other metal systems, copper is intentionally alloyed to improve its strength without, at the same time, degrading ductility and workability. Alloying degrades electrical and thermal conductivity and also changes the color—from reddish brown to yellow (with Zn) or metallic white (with Ni). For decorative parts, standard alloys of particular colors are readily available. Copper alloys can be polished and buffed to almost any desired texture and luster. They can be plated, coated with organic substances, or chemically colored to further extend the variety of available finishes.

Copper has been widely exploited by humankind over thousands of years. In the last century, intensive research led to the development of new alloys, which are available on the market (Sample 1985; Alexander 1985). Improvements in refining techniques considerably increased the purity of metals used in alloying, and adventitious impurities have been progressively replaced by minor additions of other elements in such proportions that they produce a range of complex alloys that fall outside the definitions of traditional brass and bronze (CDA, *The Copper Advantage*, n.d., 4–7).

METALLURGICAL BACKGROUND

The purpose of adding alloying elements to copper is to improve certain characteristics of the metal, such as strength, hardness, and corrosion resistance, or to effect a change in color (Joseph 2001). The final arrangement of the alloying elements with respect to the normal copper lattice determines the properties of the alloy material. Regarding strengthening (fig. 5.1a), some elements such as tin, zinc, aluminum, manganese, and silicon are usually added by solid solution to strengthen the alloy, while others (beryllium, chromium, and zirconium) act by precipitation hardening (PH) (Davis 2001b, 31). Depending on the solubility and the concentration of the alloying element, several combinations are possible, and certain properties will be preferentially improved (fig. 5.1b).

According to similarities in composition, copper alloys are conventionally grouped into six families: coppers, high-coppers, brasses, bronzes, copper-nickels, and nickel-silvers. They are briefly described below and summarized in table 5.1.

The first two families contain only minute amounts of other elements. The *coppers* family is essentially commercially pure copper. Normally soft and ductile, these alloys can be strengthened by cold working. However, exposure to elevated temperatures readily anneals the cold-worked structure. The *high-copper alloys* have minimum copper contents in excess of 94%. Alloying serves to impart greater strength, hardness, and wear resistance, but the amount of foreign element is not enough to change the face-centered cubic (FCC) structure of Cu (Davis 2001b, 31–35).

The remaining four families—brasses, bronzes, copper-nickels, and nickel-silvers—are copper alloys with copper contents lower than about 94% but not less than 50%. *Brasses* are alloys containing zinc as the principal alloying

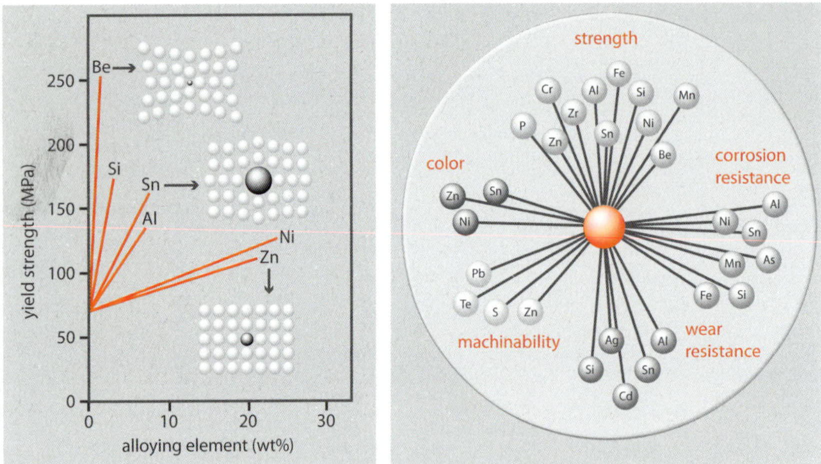

a b

Figure 5.1 | Effect of several alloying elements on the yield strength of copper: Nickel and zinc atoms are about the same size as copper atoms, but beryllium and tin atoms are very different in size, increasing strengthening (a) (adapted from Askeland and Wright 2016, fig. 10-6); alloying elements and properties (b) (adapted from CDA 2004, cover).

Table 5.1. | General classification of wrought and cast copper alloys.

Generic name		Composition	
Family	Sub family	Wrought	Cast
Coppers	–	>99% Cu	>99% Cu
High-coppers	–	>96% Cu	>94% Cu
Brasses	Brasses (leaded)	Cu-Zn (Pb)	–
	Red brasses (leaded)	–	Cu(75%–89%)-Zn-Sn (Pb)
	Yellow brasses (leaded)	–	Cu(57%–74%)-Zn-Sn (Pb)
Bronzes	Phosphor bronzes (leaded)	Cu-Sn-P (Pb)	–
	Aluminum bronzes	Cu-Al-Ni-Fe-Si-Sn	Cu-Al-Fe-Ni
	Silicon bronzes	Cu-Si-Sn	–
	Manganese bronzes (leaded)	–	Cu-Zn-Mn-Fe (Pb)
	Silicon bronzes, silicon brasses	–	Cu-Zn-Si
	Tin bronzes (leaded)	–	Cu-Sn-Zn (Pb)
	Nickel-tin bronzes	–	Cu-Ni-Sn-Zn-Pb
Copper-nickels	–	Cu-Ni-Fe	Cu-Ni-Fe
Nickel-silvers	–	Cu-Ni-Zn	Cu-Ni-Zn-Pb-Sn

Source: Data from Davis 2001a, 14.

metal, with or without secondary elements. Alloys with up to about 30% zinc have a single-phase microstructure, called the alpha phase, which is a solid solution of zinc in copper (Davis 2001b, 35–45). Brasses with more than about 37% zinc have a binary microstructure (two-phase) and are known as alpha-beta alloys (see fig. 1.15). Increasing zinc content produces stronger alloys at the expense of a moderate decrease in corrosion resistance. There are many other subfamilies, such as ordinary brasses or red, semi-red, and yellow brasses, classified according to alloy content and/or whether they have been cast or wrought.

Bronzes are copper alloys in which the major element is neither zinc nor nickel. The term originally described alloys where tin was the only or the principal element. Today the designation is used for a variety of copper alloys, including those with little or no tin. This is because they resemble "antique bronze" in both natural and weathered colors. Today the term bronze is generally used not alone but with a modifying adjective to describe the principal added element.

Phosphor bronzes, sometimes called *tin bronzes*, present improved corrosion resistance and tensile strength owing to their tin content and have an increased wear resistance and stiffness induced by the presence of phosphorus. *Aluminum bronzes* are used for their combination of high strength and excellent corrosion and wear resistance; they can be heat-treated to enhance mechanical properties for more demanding applications. *Silicon bronze* is known for its easy pouring ability, attractive golden bronze color, and superior corrosion resistance, even when submerged in liquids and chemicals.

Copper-nickels (or *cupronickels*) are alloys that utilize the complete solid solubility of copper and nickel (see fig. 1.14) to provide a range of single-phase alloys that offer excellent corrosion resistance and strength. This family also includes various precipitation-hardening alloys owing to the formation of hardening phases with third elements (CDA, *Copper-Nickel Alloys*, n.d).

Nickel-silvers are alloys that contain copper, nickel, and zinc. They present good corrosion resistance, and mechanical properties can be improved by solid solution strengthening and work hardening. Their attractive silver luster is exploited for decorative hardware, electroplated tableware, optical and photographic equipment, and musical instruments (CDA 1965).

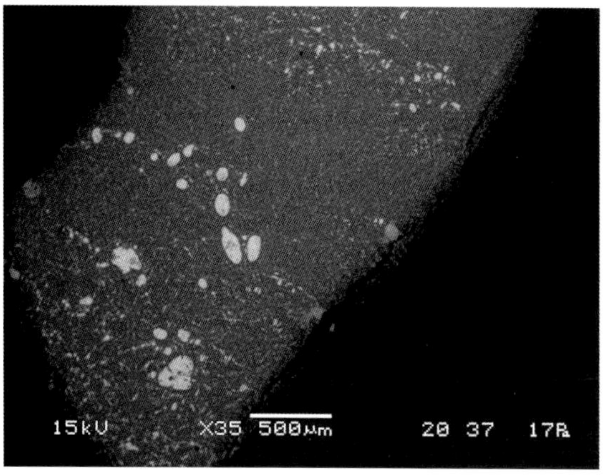

Figure 5.2 | Lead in copper alloys forms discrete microscopic pools (white in the figure), sealing pores and improving machinability (reproduced from CDA 1994, fig. I-10).

Several cast alloys have lead contents of 5% or more. Because of the low solubility of lead in copper (see fig. 1.16), it remains separated and is thus able to partially seal interdendritic pores that form during solidification (fig. 5.2). Lead also acts as a lubricant for cutting tool edges and promotes the formation of small, discontinuous chips that can easily be cleared, allowing the machining of castings to be performed at higher speeds without the aid of coolants.

DESIGNATION, PROPERTIES, AND APPLICATIONS

Copper alloys are identified by composition and temper—that is, their metallurgical condition resulting from the thermomechanical processing history. Under the Unified Numbering System for Metals and Alloys (UNS), copper and its alloys are designated by five-digit numbers preceded by the letter *C*. In the designation system, numbers from C10000 through C79999 denote wrought alloys (Tyler 1990). Cast alloys are numbered from C80000 through C99999 (Schmidt and Schmidt 1990). Compositions of casting alloys may differ from those of their wrought counterparts for various reasons. Generally, casting permits greater freedom in the use of alloying elements because it is not important to consider the effects of composition on hot- or cold-working properties (CDA 1994).

Coppers must contain at least 99.3% Cu but may include residual deoxidizers (e.g., carbon, phosphorus, boron) or minor alloying elements (table 5.2). They present

Table 5.2. | Characteristics of selected wrought and cast coppers.

Common name (UNS)	Nominal composition	Commercial forms/casting types	Typical applications
Wrought (UNS C10100–C15999)			
Oxygen-free (C10200)	99.95% Cu	F, R, W, T, P, S	Busbars, wave guides, vacuum tubes
Tough-pitch copper (C11000)	99.90% Cu, O	F, R, W, T, P, S	Electrical wire and cable; roofing and architectural trim
Phosphorus deoxidized copper (C12200)	99.90% Cu, P	F, R, T, P	Plumbing pipe and tubing; condenser, steam, and water lines
Cast (C80100–C81200)			
Oxygen-free copper (C80100)	99.95% Cu, Ag	C, T, I, M, P, S	Terminals, connectors, and other corrosion- and oxidation-resistant applications
High-conductivity copper (C81100)	99.7% Cu, Ag	C, T, I, M, P, S	

Source: Data from Tyler and Black 1990.
Note: Commercial forms: F = flat, R = rod, W = wire, P = pipe, S = shapes, T = tube; casting types: C = centrifugal, T = continuous, D = die, I = investment, M = permanent mold, P = plaster, S = sand.

Table 5.3. | Characteristics of selected wrought and cast high-coppers.

Common name (UNS)	Nominal composition	Commercial forms/ casting types	Properties and typical applications
Wrought (UNS C16200–C19900)			
Beryllium-copper (C17000)	99.5% Cu, Be, Co	F, R	Precipitation hardenable; fasteners, springs, valves, welding equipment
Cadmium-copper (C16200)	99.0% Cu, Cd	F, R, W	Improved resistance to thermal softening and wear resistance
Chromium-copper (C18200)	99.5% Cu, Cr	F, R, W, T, S	Precipitation hardenable, high strength, and thermal softening resistance; electrical conductivity higher than 80% IACS
Cast (C81400–C82800)			
Chromium copper (C81500)	98.0% Cu, Cr	C, T, I, M, P, S	Electrical contacts, resistance welding electrodes, dies, bearings
Beryllium copper (C82200)	96.5% Cu, Ni, Be	C, T, I, M, P, S	

Source: Data from Tyler and Black 1990.
Note: Commercial forms: F = flat, R = rod, W = wire, P = pipe, S = shapes, T = tube; casting types: C = centrifugal, T = continuous, D = die, I = investment, M = permanent mold, P = plaster, S = sand.

superior electrical conductivity (100% IACS), have excellent hot and cold workability, and are resistant to atmospheric and aqueous corrosion.

High-coppers contain ~94% Cu and small amounts of various alloying elements such as beryllium, cadmium, chromium, iron, nickel, cobalt, and tin to yield higher strength and thermal stability while retaining sufficient electrical conductivity (table 5.3).

Common brasses are copper alloys in which zinc is the principal alloying element (table 5.4). They exhibit good strength and ductility and are easily cold-worked. Brass coloration ranges from red to pale yellow, depending on zinc content. Brasses containing high zinc content, such as Muntz metal, have high strength and lower ductility at room temperature than alloys with less zinc. There are also miscellaneous copper-zinc alloys, whose mechanical properties are further enhanced by additions of Mn, Fe, Sn, Al, Si, and/or Co. Free-cutting brass contains fine particles of lead that promote the formation of short and friable shavings during machining, reduce wear on the cutting

Table 5.4. | Characteristics of selected wrought and cast copper-zinc alloys.

Common name (UNS)	Nominal composition	Commercial forms/ casting types	Typical applications
Wrought brasses (UNS C20100–C48600)			
Gilding, 95% (C21000)	95.0% Cu, Zn	F, W	Coins, medals, jewelry base for gold plate
Commercial bronze, 90% (C22000)	90.0% Cu, Zn	F, R, W, T	Screws, rivets, lipstick cases
Jewelry bronze, 87.5% (C22600)	87.5% Cu, Zn	F, W	Chain, fasteners, costume jewelry
Red brass, 85% (C23000)	85.0% Cu, Zn	F, W, T, P	Condenser and heat exchanger tubing, fire extinguishers
Low brass, 80% (C24000)	80.0% Cu, Zn	F, W	Battery caps, musical instruments, clock dials
Cartridge brass, 70% (C26000)	70.0% Cu, Zn	F, R, W, T	Ammunition components, locks, plumbing accessories
Yellow brass (C27000)	65.0% Cu, Zn	F, R, W	Same use as cartridge except for ammunition
Muntz metal (C28000)	60.0% Cu, 40.0 Zn	F, R, T	Large nuts and bolts, architectural panel sheets, brazing rods
Architectural bronze (C38500)	57.0% Cu, Zn, Pb	R, S	Architectural extrusions, store fronts, trim, lock bodies
Admiralty metal (C44300)	71.0% Cu, Zn, Sn	F, W, T	Heat exchanger and condenser tubes
Naval brass (C46400–C46700)	60.0% Cu, Zn, Sn	F, R, T, S	Propeller shafts, marine hardware
Other Cu-Zn alloys (C66300–C69710)			
Manganese bronze A (C67500)	58.5% Cu, Zn, Fe, Sn, Mn	F, R	Pump rods, shafting balls, brazing alloy
Aluminum bronze, arsenical (C68700)	77.5% Cu, Zn, Al, As	T	Heat exchanger and condenser tubes
Cast brasses (C83300–C87900 and C89320–C89940)			
Red and semired (leaded) brasses (C83300–C84800)	Cu-Sn-Zn-Pb	S and others depending on the alloy	Electrical hardware, cable connectors; general-purpose alloys for plumbing and hardware
Yellow (leaded) brasses (C85200–C85800)	Cu-Zn (Pb)	Depends on the alloy	Architectural trim, decorative hardware
Manganese bronze (high-strength yellow brasses) (C86100–C86800)	Cu-Zn-Mn-Fe (Pb)	S and others depending on the alloy	Gears, levers, valve and pump components for freshwater and seawater service
Silicon brasses/bronzes (C87200–C87900)	Cu-Zn-Si	Depends on the alloy	Bearings, gears, plumbing goods, intricately shaped valve and pump components
Copper-bismuth and Cu-Bi-Se brasses (C89510–C89520)	Cu-Sn-Zn-Bi-Se	Depends on the alloy	Plumbing fixtures, valves, and fittings for potable water and food processing

Source: Data from Tyler and Black 1990.
Note: Commercial forms: F = flat, R = rod, W = wire, P = pipe, S = shapes, T = tube; casting types: C = centrifugal, T = continuous, D = die, I = investment, M = permanent mold, P = plaster, S = sand.

tool, and permit parts to be processed rapidly on automatic machines. Brasses have excellent castability, relatively low cost, and a favorable combination of strength and corrosion resistance.

Bronzes are commonly named by their principal alloying element (Davis 2001b, 45–53) (table 5.5). *Phosphor bronzes* (C50100–C54400) are also commonly referred to as *tin bronzes*. They have superior spring qualities, high fatigue resistance, excellent formability and solderability, and high corrosion resistance. The addition of lead improves machinability (free cutting).

Aluminum bronzes (C60800–C64210) contain up to 14% aluminum and frequently other elements such as nickel, iron, manganese, and silicon. Varying the proportions of the other elements results in a range of strong, tough alloys with excellent resistance to corrosion and wear (CDA 1998, 42). They are readily weldable and can be machined or ground with good lubrication.

Silicon bronzes (C64700–C66100) are produced in relatively low volumes for products such as high-strength fasteners and marine hardware. These alloys, with their excellent weldability and good corrosion resistance, are becoming the alloy of choice for a quality casting (Kipper 1996, 23).

Copper-nickel alloys (or cupronickels) were developed in the 1930s and contain up to 30% nickel with other elements (table 5.6). The addition of nickel to copper improves strength

Table 5.5. | Characteristics of selected wrought and cast bronzes.

Common name (UNS)	Nominal composition	Commercial forms/ casting types	Typical applications
Wrought phosphor bronzes (C50100–C54400)			
Phosphor bronze, 5% A (C51000)	95.0% Cu, 5.0% Sn, trace P	F, R, W, T	Fasteners, lock washers, chemical hardware, welding rod
Phosphor bronze, 10% D (C52400)	90.0% Cu, 10.0% Sn, trace P	F, R, W	Heavy bars and plats for severe compression; articles requiring spring qualities, good wear, and corrosion resistance
Free-cutting phosphor bronze (C54400)	88.0% Cu, 4.0% Pb, 4.0% Zn, 4.0% Sn	F, R	Bearings, gears, shafts, valve parts
Wrought aluminum bronzes (C60800–C64210)			
Aluminum bronze, 5% (C60800)	95.0% Cu, 5.0% Al	T	Condenser, distiller tubes, heat exchanger tubes
Aluminum bronze, D (C61400)	91.0% Cu, 7.0% Al, 2.0% Fe	F, R, W, T, P, S	Corrosion-resistant vessels and tanks, machine parts, nuts and bolts
(C61500)	90.0% Cu, 8.0% Al, 2.0% Ni	F	Decorative metal trim, interior furnishings, and items requiring high tarnish resistance
Wrought silicon bronzes (C64700–C66100)			
Low-silicon bronze, B (C65100)	98.5% Cu, 1.5% Si	R, W, T	Hydraulic pressure lines, marine hardware, welding rod, heat exchanger tubing
High-silicon bronze, A (C65500)	97.0% Cu, 3.0% Si	F, R, W, T	Similar to C65100, including propeller shafts
Cast bronzes (C90200–C95900)			
Tin and leaded tin bronzes (C90200–C94500)	Cu-Sn-Zn/Pb	C, S, and others depending on the alloy	Gears, marine fittings, pump components Examples: Navy G bronze (C90300), gunmetal (C90500), ring metal (C92800), bearing bronze (C93200), phosphor bronze (C94500)
Nickel tin bronzes (C94700–C94800)	Cu-Ni-Sn-Zn-Pb	M, S, and others depending on the alloy	Impellers, screw conveyors, machinery components
Aluminum bronzes (C95200–C95900)	Cu-Al-Fe-Ni	C, T, M, P, S	Heavy-duty mechanical and structural products including gears, valve guides, and seats

Source: Data from Tyler and Black 1990.
Note: Commercial forms: F = flat, R = rod, W = wire, P = pipe, S = shapes, T = tube; casting types: C = centrifugal, T = continuous, D = die, I = investment, M = permanent mold, P = plaster, S = sand.

and corrosion resistance while retaining ductility. These are among the most corrosion resistant and thermally stable of all copper alloys and are virtually immune to stress-corrosion cracking (SCC).

Nickel-silvers contain nickel and zinc in various amounts (table 5.7). The development of modern nickel-silvers arose in the early nineteenth century from the need for an alloy closely resembling silver yet much less expensive (CDA 1965). Around 1824–25, investigators in Germany and France directed their efforts toward improving paktong, the ancient Chinese copper-nickel-zinc alloy (CDA 1965; Cowden 2014, 25). When the alloys were first produced in Britain a year or two later, they were called German silver; they are also known as *maillechort* in France and *alpaca* in South America. They combine good corrosion resistance with moderately high strength, which accounts for their wide use in food and beverage handling equipment. Their attractive silver luster is exploited for decorative hardware, electroplated tableware, optical and photographic equipment, jewelry, and musical instruments. Their color ranges from yellow to silvery white (Pieper 2015).

There are also a number of special cast alloys (C99300–C99750) with unique characteristics—such as Incramet 800 (UNS C99300)—that have been developed for glass processing. They present high oxidation resistance and good thermal fatigue resistance. Other alloys, including

Table 5.6. | Characteristics of selected wrought and cast copper-nickel alloys.

Common name (UNS)	Nominal composition	Commercial forms/ casting types	Typical applications
Wrought copper-nickels (C70100–C72950)			
Copper-nickel, 10% (C70600)	88.7% Cu, Ni, Fe	F, T	Condenser, evaporator tubing, saltwater piping, vehicle hydraulic systems, coinage
Copper-nickel, 30% (C71500)	70.0% Cu, Ni	F, R, T	
Cast copper-nickels (C96200–C96950)			
Copper-nickel (C96200–C96950)	Cu-Ni-Fe Cu-Ni-Sn (spinodal)	C, S, and others depending on the alloy	Marine valve components, oil and gas, aerospace

Source: Data from Tyler and Black 1990.
Note: Commercial forms: F = flat, R = rod, W = wire, P = pipe, S = shapes, T = tube; casting types: C = centrifugal, T = continuous, D = die, I = investment, M = permanent mold, P = plaster, S = sand.

Table 5.7. | Characteristics of selected nickel-silver alloys.

Common name (UNS)	Nominal composition	Commercial forms/ casting types	Typical applications
Wrought (C73500–C79830)			
Nickel-silver, 65-10 (C74500)	65.0% Cu, Zn, Ni	F, W	Rivets, screws, optical parts
Nickel-silver, 65-18 (C75200)	65.0% Cu, Ni, Zn	F, R, W	Table flatware, zippers, camera parts, base for silver plate, costume jewelry
Cast (C97300–C97800)			
Nickel-silver (C97300–C97800)	Cu-Sn-Pb-Zn-Ni	S, I M, C depending of the alloy	Dairy metal (C97600)

Source: Data from Tyler and Black 1990.
Note: Commercial forms: F = flat, R = rod, W = wire, P = pipe, S = shapes, T = tube; casting types: C = centrifugal, T = continuous, D = die, I = investment, M = permanent mold, P = plaster, S = sand.

the Incramute 1 (UNS C99600), are used in musical instrument components; they have good damping characteristics and attenuate noise well.

Besides the composition, copper alloys are also designated by their metallurgical condition. Temper designations are based on the sequences of treatments used to produce the part (Davis 2001a, 14–15). Alloys are assigned an alphanumeric code defined by a standard designation system, ASTM B601, used for all forms of wrought and cast copper and copper alloys (CDA, *The Copper Advantage*, n.d., 12–14). The basic designation consists of letters, as presented in table 5.8.

Where required, letters are followed by one or more digits corresponding to specific sequences of basic treatments—H04 for hard, H08 for spring, HR08 for spring and stress-relieved, M01 for as-sand-cast, M20 for as-hot-rolled, and so on. Temper designations only imply specific mechanical properties when used in association with a particular alloy, product form, and size.

Cold-worked or heat-treated alloys are said to have a harder temper, and those annealed are said to have a softer temper. Harder tempers present higher strength and hardness but at the expense of reduced ductility. Machinability, as measured by tool wear rates, usually deteriorates in proportion to the hardness of the alloy. Electrical and thermal conductivity vary with the degree of temper, but the nature and extent of the effect depends strongly on the type of alloy. Chemical properties such as corrosion resistance are not greatly affected by temper, although residual cold-work-induced tensile stresses render some copper alloys more susceptible to stress-corrosion cracking than when the metals are in the annealed state.

Table 5.8. | Temper designations for copper and copper alloys.

Designation	Material condition
H	Cold-worked
HR	Cold-worked and stress-relieved
M	As-manufactured
O	Annealed
TB	Solution-treated
TD	Solution-treated and cold-worked
TF	Solution-treated and precipitation-hardened
TM	Mill-hardened
TQ	Quench-hardened

Source: Data from Davis 2001a, 26–27.

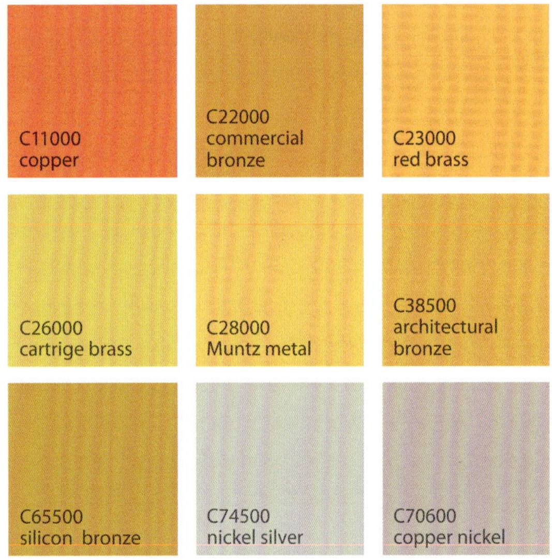

Figure 5.3 | Various natural copper alloy colorations (reproduced from CDA, *The Copper Advantage*, n.d., fig. 6).

Regarding appearance, the color of copper alloys is primarily given by their chemical composition. Unalloyed coppers have a red tone. The addition of other elements causes a change toward yellow, bronze, silver, or gray (fig. 5.3).

The original color can be modified either by the development of natural patinas upon exposure to air or by chemical means intended to hasten natural weathering. For a given alloy composition, the result varies with the products used, surface preparation, and skill, as well as environmental factors. Before final finishing operations, it is usually necessary to perform intermediate steps, such as immersing the metal in an acidic bath to remove oxides formed during annealing or welding or to produce matte or bright surfaces (Brock 1994). Once adequately prepared, the metal surface can be further naturally or chemically converted to a stable, protective compound (fig. 5.4).

The characteristic blue-green patina of natural weathered copper is a direct consequence of the mild corrosive attack by airborne sulfur compounds. Natural patinas generally consist of mixtures of basic copper carbonate and basic

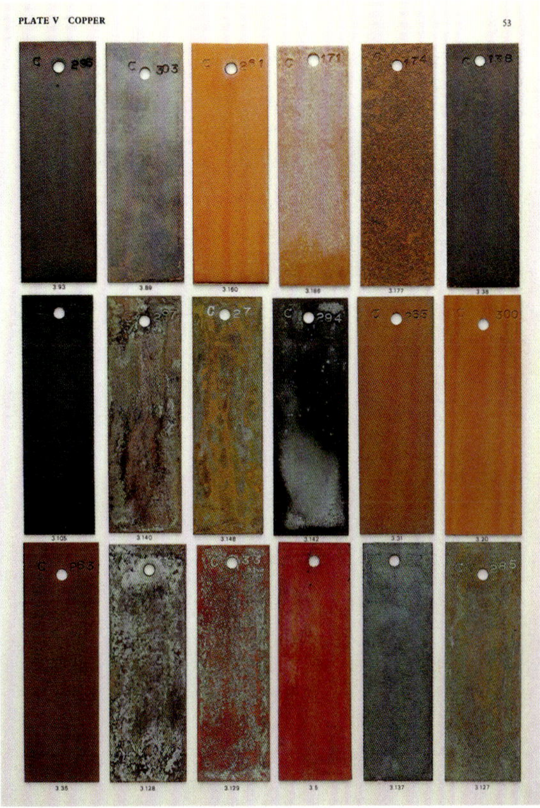

Figure 5.4 | Possible patinas on copper alloys (reproduced from Hughes and Rowe 1993, plate V).

copper sulfate; the exact composition strongly depends on the type and concentration of atmospheric constituents. As natural weathering proceeds, the metal exposed to the atmosphere changes in hue from the natural salmon pink color; it passes through a series of brown shades (initial oxide film) that finally become a dull gray or black (sulfide films), from which the ultimate blue-green or gray-green (basic copper sulfate) patinas emerge (fig. 5.5). These films are quite thin and, while adherent, have relatively low abrasion resistance. Once completely formed, the sulfate patina is rather resistant to atmospheric corrosion. Natural patinas are far from uniform, because factors such as temperature, drainage, insolation, and surface slope affect formation rate and appearance (CDA, *Copper in Architecture*, n.d., 1.2.1–1.2.5).

Chemical conversion coatings (also known as chemical patinas) are developed in a wide range of colored finishes by applying reagents through brushing, sponging, stippling, or spraying, often with heat (Hughes and Rowe 1993; NAAMM and NOMMA 2006; Fishlock 1962). The wide range of colors and shades that may be achieved depends upon a number of variables, such as reagents, surface preparation, time, and temperature, and creating these is largely a matter of skill. Several conversion treatments are in general use. The most common colors are referred to as brown, or statuary, finishes for bronze and as green, or patina, finishes for copper. Figure 5.6 shows some examples.

Statuary finishes are based on a mixture of copper oxides and sulfides. They are commonly produced from solutions of polysulfides, usually called "oxidizing liquid" and sold in concentrated form under a number of trade names. Statuary finishes are commonly hand-applied and hand-rubbed periodically with wax or oils to maintain a stable appearance (CDA, *How to Apply Statuary and Patina Finishes*, n.d.). Their brownish colors range from light to dark, depending on the copper alloy, surface texture, and application conditions of the coloring solutions.

Patina finishes are generally produced by the action of acid-chloride or acid-sulfate treatments, and they aim to

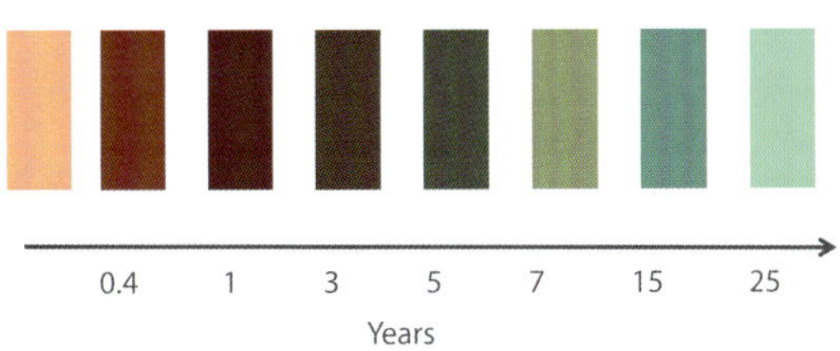

Figure 5.5 | Natural weathering color chart showing typical changes for copper in most industrial environments (reproduced from CDA and Canadian Copper and Brass Development Association, n.d., fig. III-7).

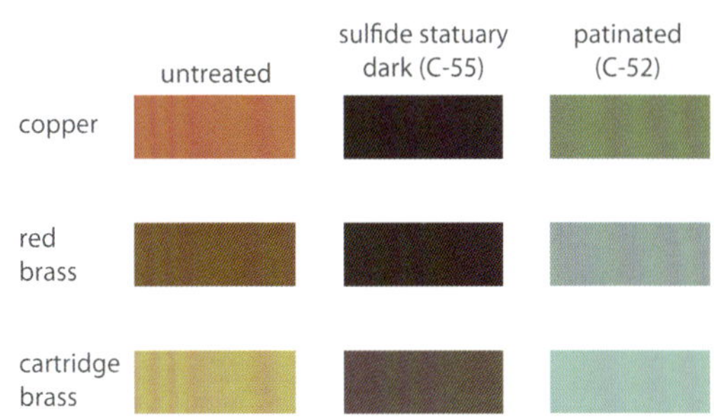

Figure 5.6 | Chemical finishes: The result depends on the alloy type, the chemical solution, and the application method (reproduced from CDA and Canadian Copper and Brass Development Association, n.d., fig. VI-15).

create the desired finish rapidly (CDA, *How to Apply Statuary and Patina Finishes*, n.d.). Since there is a large number of variables involved—including the manner in which the solutions are applied, the weather conditions under which the treatment is carried out, and the climate to which the treated surfaces are exposed—such chemically induced patinas are prone to lack of adhesion, staining of adjacent materials, and inadequate color uniformity over large surface areas. These potential shortcomings should be considered when attempting to artificially duplicate a weathering process.

DETERIORATION

Copper and copper alloys are widely used in many environments and applications because of their excellent corrosion resistance, which is coupled with combinations of other desirable properties. At ambient temperatures, the reaction product predominantly responsible for protection is cuprous oxide (Cu_2O). This film is adherent, and the corrosion rate is significantly reduced by the presence of cations such as Al, Zn, Sn, Fe, and Ni coming from alloying elements. Copper alloys corrode at negligible rates in unpolluted air, water, and deaerated nonoxidizing acids. They are, however, susceptible to rapid attack in oxidizing acids, sulfur, ammonia, and some of their compounds. Resistance to acid solution depends mainly on the severity of the oxidizing conditions in the solution (Tyler and Black 1990; Cohen 2005).

Regarding galvanic coupling effects, copper alloys are generally cathodic to other common structural metals. When put in contact with steel or aluminum, these metals corrode with an increased rate, while the copper-alloy corrosion rate decreases. However, when coupled with high-nickel alloys or titanium, copper metals usually corrode preferentially. With respect to stainless steel, copper alloys can be anodic or cathodic, depending on exposure conditions and the passivity of the stainless steel. Galvanic corrosion between two different copper alloys is rare, because the potential difference is very small. Alloying with zinc or aluminum shifts the potential toward the anodic end of the galvanic series in seawater; the addition of tin and nickel moves the potential toward the cathodic end.

Concerning localized corrosion, although copper alloys generally do not corrode by pitting, the corroded surface usually shows a tendency toward nonuniformity (Roberge 2008). High-copper alloys and silicon bronzes are most prone to pitting, while copper-nickels and tin bronzes tend to have intermediate resistance. The aluminum bronzes and low-zinc brasses present a high resistance to pitting. Moreover, for most copper alloys, the attack is located outside but immediately adjacent to the crevice owing to the formation of metal ion concentration cells. Local cell action similar to crevice attack may also result from the presence of foreign objects or debris, such as dirt, vegetation, rust, or the uneven accumulation of corrosion products on the surface. Localized pitting reduces load-carrying capacity and increases stress concentration by creating depressions or holes in the metal.

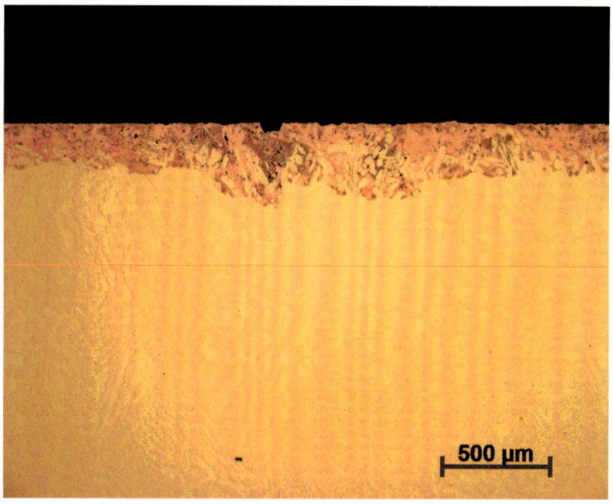

Figure 5.7 | Cross section of a Cu-42Zn alloy exhibiting dezincification; notice the copper-rich layer at the surface. Photo: Johann Wilhelm Erning

Dealloying, or selective corrosion, is observed in copper alloys with high alloying element content. Thus, dezincification (selective removal of zinc leaving a porous layer of copper and copper oxide) is the usual form of corrosion for brasses with more than 15% Zn in prolonged contact with waters high in oxygen and carbon dioxide (fig. 5.7). For biphasic alloys, dezincification is more severe. Some alloying elements, such as Sn, P, As, and Sb, tend to inhibit dezincification. Cases of dealloying have been observed in copper-aluminum alloys, particularly those having more than 8% Al.

Intergranular corrosion, which penetrates the metal along grain boundaries, is observed in copper alloys, of which the most susceptible are Muntz metal, admiralty metal, aluminum brasses, and silicon bronzes. Stress corrosion is also observed when the alloy is under applied or residual stresses in a corrosive environment, especially in the presence of

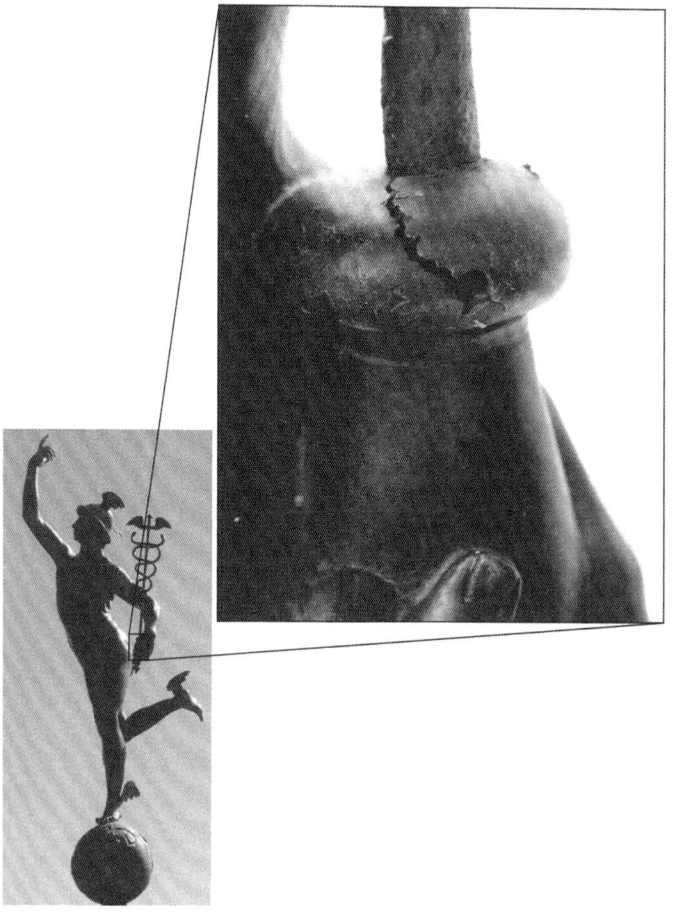

Figure 5.8 | Stress-corrosion crack in a bronze monument: Galvanic corrosion of the internal iron rod in contact with the bronze in a moist environment causes voluminous rust buildup, generating high stresses and inducing SCC (reproduced from Roberge 2012, fig. 6.46).

oxygen and moisture (fig. 5.8). Brasses containing 20%–40% Zn are highly susceptible to stress-corrosion cracking.

Impingement attack occurs where liquids or gases impinge on metal surfaces at high velocities. Rapidly moving turbulent water can strip away the protective films from copper alloys, allowing the metal to corrode at a more rapid rate.

CHARACTERIZATION

Most copper alloys have a characteristic red-yellow color that facilitates their visual distinction from other metallic alloys. It is, however, very complicated to differentiate among all possible existing copper alloys. It is no longer a question of whether one is a "bronze" or a "brass" because there are currently hundreds of copper alloys. As pointed out in chapter 3, a complete characterization of contemporary alloys requires not only a careful quantitative analysis (since the minor alloying elements play a key role) but also a structural characterization that attests to the metallurgical history. In any case, a preliminary visual examination of the part (color, use/application, etc.) (table 5.9) can be useful in determining the type of alloy, thus allowing for the selection of the suitable analytical tool to be used for further study.

Regarding the conservation condition, copper alloys are usually covered by colored surface products, which are smooth and tightly adherent. The presence of small powdery spots, generally over the entire surface, indicates that the corrosion can be locally activated (see chap. 2, "Pitting") (National Park Service 2002, 04).

Table 5.9. | Color matching with fixture forms of compatible copper alloys.

Color	Castings	Fasteners	Rod and wire	Extrusions	Sheet and plate
Copper-red	Copper (99.9% min.)	C11000 C65100	C11000 C12500	C11000 C12500	C11000 C12200 C12500
Bronze-gold	C83400	C65100	C22000	C31400	C22000
Yellow-gold	C85200 C85300	C26000 C36000 C46400 C48600	C26000	C26000	C26000
Pinkish gray	C96200	C70600	C70600	C70600	C70600
Warm silver	C97300	C74500	C74500 C75200	C79600	C74500 C75200

Source: Data from CDA, *The Copper Advantage*, n.d., 26.

CONSERVATION-RELATED LITERATURE

Publications on the conservation of contemporary copper alloys are mainly related to sculptures. There are many case studies of artifacts from well-known artists, such as Daniel Spoerri (Boll and Scheiblauer 2013), Richard Lippold (Flotte and Dommermuth 2000), Francesco Messina (Marcelli 2004), Alberto Viani (Longega, Driussi, and Morabito 2004), and Donald Judd (Nagy 1999), as well as those belonging to the Stark Collection at the Getty Center (Considine et al. 2010) and others presenting stabilization problems (Brunnert 2003). Results of research investigating artists' materials (Nagy 2004; Nagy, Adamsons, and Moomaw 2013), evaluating corrosion evolution under given conditions (Chiavari et al. 2009; Laguzzi, Luvidi, and Brunoro 2001; Selwyn et al. 1996; Letardi et al. 2016), and correlating composition and aesthetic aspects to given foundries (Young et al. 2009; Day et al. 2010) have also been published. Some review papers discuss aesthetic and technical considerations in the evaluation, care, and maintenance of sculptures (Boulton 2007; Singer and Fletcher 2004; Conner 2006), while a few of them address architectural applications, such as the use of nickel-silvers (Cowden 2014) and materials employed for roofing and decorative elements of facades (Gayle and Look 1992). Scientific instruments have been studied in a master's thesis (Geffroy 2004).

REFERENCES AND BIBLIOGRAPHY

Alexander, B. O. 1985. "Copper 5: Other Alloys." *Historical Metallurgy* 19 (1): 28–30.

Askeland, D. R., and W. J. Wright. 2016. *The Science and Engineering of Materials*. Boston: Cengage Learning.

Becker, A., U. Ruhrberg, and T. Jentzch. 2015. "Pitting Corrosion of Copper: Current Damages in Drinking Water Systems in Germany." In *EUROCORR 2015, Graz 06–10 September 2015*, paper 564. Northampton, UK: European Federation of Corrosion.

Boll, B., and C. Scheiblauer. 2013. "*Die Prillwitzer Idole* von Daniel Spoerri: Konservierung von zeitgenössischen Bronzeskulpturen." *Beiträge zur Erhaltung von Kunst- und Kulturgut*, no. 2:110–15.

Boulton, A. 2007. "The Making of Matisse's Bronzes." In *Matisse: Painter as Sculptor*, edited by D. Kosinsky, J. M. Fisher, and S. Nash, 73–97. New Haven, CT: Yale University Press.

Brock, A. J. 1994. "Surface Engineering of Copper and Copper Alloys." In *Surface Engineering*, vol. 5, 805–18. ASM Handbook. Materials Park, OH: ASM International.

Brunnert, S. 2003. "Instabile Plastiken aus Kupfer und Kupferbasislegierungen im Freien. Teil 1, Lösungsansätze für zeitgenössische Plastiken aus Bronze." *Restauro* 109 (7): 484–89.

CDA [Copper Development Association]. 1965. *The Nickel Silvers: Design Data and Applications*. Hemel Hempstead, UK: Copper Development Association. *http://www.copper.co.za/downloads/library/71-The-Nickel-Silvers.pdf.*

———. 1994. *Copper Casting Alloys*. New York: Copper Development Association. *https://www.copper.org/publications/pub_list/pdf/7014.pdf.*

———. 1998. *Copper–The Vital Metal*. Hemel Hempstead, UK: Copper Development Association. *https://copperalliance.org.uk/resources/pub-121-copper-vital-metal/.*

———. 2004. *Copper and Copper Alloys: Compositions, Applications and Properties*. Hemel Hempstead, UK: Copper Development Association. *http://copperalliance.org.uk/docs/librariesprovider5/resources/pub-120-copper-and-copper-alloys.pdf.*

———. n.d. *The Copper Advantage: A Guide to Working with Copper and Copper Alloys*. New York: Copper Development Association. *https://www.copper.org/publications/pub_list/pdf/a1360.pdf.*

———. n.d. *Copper in Architecture: A Comprehensive Compilation of Designs, Details and Specifications*. New York: Copper Development Association. *https://www.copper.org/publications/pub_list/pdf/A4050-Architectural-Handbook.pdf.*

———. n.d. *Copper-Nickel Alloys: Properties, Processing, Applications*. New York: Copper Development Association. *https://www.copper.org/applications/marine/cuni/properties/DKI_booklet.html.*

———. n.d. *How to Apply Statuary and Patina Finishes*. New York: Copper Development Association. *https://www.copper.org/publications/pub_list/pdf/a1081.pdf.*

CDA [Copper Development Association] and Canadian Copper and Brass Development Association. n.d. *Copper Brass Bronze Design Handbook: Architectural Applications*. New York: Copper Development Association; North York, ON: Canadian Copper and Brass Development Association. *https://www.copper.org/publications/pub_list/pdf/A4039-ArchitecturalApplications.pdf.*

Chiavari, C., E. Bernardi, F. Ospitali, L. Robbiola, C. Martini, and L. Morselli. 2009. "La Corrosione atmosferica dei monumenti in bronzo: Prove di invecchiamento artificiale." *La Metallurgia italiana* 101 (5): 45–54.

Cohen, A. 2005. "Copper (and Alloys)." In *Corrosion Tests and Standards: Application and Interpretation*, 2nd ed., edited by R. Baboian, 466–75. West Conshohocken, PA: ASTM International.

Conner, J. 2006. "Introduction to Cast Comparisons." In *Captured Motion: The Sculpture of Harriet Whitney Frishmuth*, edited by J. Conner, L. R. Lehmbeck, and T. Tolles, 101–26. New York: Hohmann Holdings.

Considine, B., J. Wolfe, K. Posner, and M. Bouchard. 2010. *Conserving Outdoor Sculpture: The Stark Collection at the Getty Center*. Los Angeles: Getty Conservation Institute.

Cowden, A. B. 2014. "Nickel Silver." In *Twentieth-Century Building Materials: History and Conservation*, edited by T. C. Jester, 25–30. Los Angeles: Getty Conservation Institute.

Davis, J. R. 2001a. *Copper and Copper Alloys,* edited by J. R. Davis. ASM Specialty Handbook. Materials Park, OH: ASM International.

———. 2001b. "Physical Metallurgy: Heat Treatment, Structure and Properties." In *Copper and Copper Alloys,* edited by J. R. Davis, 31–53. ASM Specialty Handbook. Materials Park, OH: ASM International.

Day, J., J. Stenger, K. Eremin, N. Khandekar, and V. Budny. 2010. "Gaston Lachaise's Bronze Sculptures in the Fogg Museum." *Journal of the American Institute for Conservation* 49 (1): 1–26.

Fishlock, D. 1962. *Metal Coloring*. Teddington, UK: R. Draper.

Flotte, J., and J. Dommermuth. 2000. "Richard Lippold's *The Sun*: Investigation and Treatment Considerations." In *Gilded Metals: History, Technology and Conservation*, edited by T. Drayman-Weisser, 329–36. London: Archetype Publications.

Gayle, M., and D. W. Look. 1992. "A Historical Survey of Metals: Copper and Copper Alloys." In *Metals in America's Historic Buildings: Uses and Preservation Treatments*, edited by M. Gayle and J. C. Waite, 22–34. Washington, DC: National Park Service.

Geffroy, Anne-Marie. 2004. "Conservation-restauration d'un instrument scientifique: Un théodolite de la fin du XIXème siècle du Musée des Arts et Métiers: Étude comparative du comportement au vieillissement artificiel de trois vernis sur laiton." MSc diss., Institut National du Patrimoine.

Hughes, R., and M. Rowe. 1993. *The Colouring, Bronzing and Patination of Metals: A Manual for Fine Metalworkers, Sculptors and Designers*. London: Thames and Hudson.

Joseph, G. 2001. "Metallurgy and Properties." In *Copper: Its Trade, Manufacture, Use, and Environmental Status*, edited by K. J. A. Kundig, 45–140. Materials Park, OH: ASM International.

Kipper, P. V. 1996. *The Care of Bronze Sculpture: Recommended Maintenance Programs for the Collector*. Coveland, CO: Path Publications.

Laguzzi, G., L. Luvidi, and G. Brunoro. 2001. "Atmospheric Corrosion of B6 Bronze Evaluated by the Thin Layer Activation Technique." *Corrosion Science* 43 (4): 747–53.

Letardi, P., B. Ramirez-Barat, M. Albini, P. Traverso, E. Cano, and E. Joseph. 2016. "Copper Alloys and Weathering Steel Used in Outdoor Monuments: Weathering in an Urban-Marine Environment." In *Metal 2016: Proceedings of the Interim Meeting of the ICOM-CC Metals Working Group, September 26–30, New Delhi, India*, edited by R. Menon, C. Chemello, and A. Pandya, 320–28. Paris: ICOM-CC.

Longega, G., G. Driussi, and Z. Morabito. 2004. "La Statua di Alberto Viani nel negozio Olivetti di Carlo Scarpa." In *Architettura e materiali del novecento: Conservazione, restauro, manutenzione: Atti del convegno di studi, Bressanone, 13–16 luglio 2004*, edited by G. Biscontin and G. Driussi, 1271–78. Scienza e beni culturali 20. Marghera-Venezia, Italy: Arcadia Ricerche.

Marcelli, I. M. 2004. "Il Restauro del 'Cavallo Morente' di Francesco Messina." In *Monumenti in bronzo all'aperto: Esperienze di conservazione a confront,* edited by P. Letardi, I. Trentin, and G. Cutugno, 155–62. Arte e restauro. Florence: Nardini Editore.

NAAMM (National Association of Architectural Metal Manufacturers) and NOMMA (National Ornamental and Miscellaneous Metals Association). 2006. "Finishes for the Copper Alloys." In *Metal Finishes Manual for Architectural and Metal Products*, 2-1 - 2-16. AMP 500-06. Chicago: National Association of Architectural Metal Manufacturers; McDonough, GA: National Ornamental and Miscellaneous Metals Association.

Nagy, E. 1999. "Corrosion Intercept® Tent Packing and Handling System for Donald Judd's Brass and Copper Sculptures: Experiences with Packing and Handling Methods." *AIC Objects Specialty Group Postprints* 6:11–35.

——. 2004. "The Maladies of Mill-Produced Metal in Modern Art." In *Modern Art, New Museums: Contributions to the Bilbao Congress, 13–17 September 2004*, edited by A. Roy and P. Smith, 67–72. London: International Institute for Conservation of Historic and Artistic Works.

Nagy, E., K. Adamsons, and K. Moomaw. 2013. "Disfiguring Organic Residues on Industrially Produced Sheet Metal Coupons Simulating Copper and Brass Works of Art by Donald Judd: Attenuated Total Reflectance Fourier Transform Infrared Spectroscopy Analysis and Treatment Recommendations." *Studies in Conservation* 58 (3): 245–55.

National Park Service. 2002. "Appendix O: Curatorial Care of Metal Objects." In *Museum Collections*. Part 1 of *The Museum Handbook*. Washington, DC: National Park Service.

Pieper, R. 2015. "The 'White Metals' of Early-Twentieth-Century American Architecture." Special Issue on Modern Metal Finishes, *APT Bulletin* 46 (1): 23–28.

Roberge, P. R. 2008. "Recognizing the Forms of Corrosion." In *Corrosion Engineering: Principles and Practice*, 147–84. New York: McGraw-Hill.

——. 2012. *Handbook of Corrosion Engineering*. 2nd ed. New York: McGraw-Hill.

Sample, S. G. 1985. "Copper 2: Brasses, Bronzes and Nickel Silver." *Historical Metallurgy* 19 (1): 12–16.

Schmidt, R. F., and D. G. Schmidt. 1990. "Selection and Application of Copper Alloy Castings." In *Properties and Selection: Nonferrous Alloys and Special-Purpose Materials*, vol. 2, 346–55. ASM Handbook. Materials Park, OH: ASM International.

Selwyn, L. S., N. E. Binnie, J. Poitras, M. E. Laver, and D. A. Downham. 1996. "Outdoor Bronze Statues: Analysis of Metal and Surface Samples." *Studies in Conservation* 41 (4): 205–28.

Singer, M. C., and V. Fletcher. 2004. "The Conservation of Arp's Bronzes: Preserving the Sculpture's History or the Artist's Intent?" In *Modern Art, New Museums: Contributions to the Bilbao Congress, 13–17 September 2004*, edited by A. Roy and P. Smith, 73–76. London: International Institute for Conservation of Historic and Artistic Works.

Tyler, D. E. 1990. "Wrought Copper and Copper Alloy Products." In *Properties and Selection: Nonferrous Alloys and Special-Purpose Materials*, vol. 2, 241–64. ASM Handbook. Materials Park, OH: ASM International.

Tyler, D. E., and W. T. Black. 1990. "Introduction to Copper and Copper Alloys." In *Properties and Selection: Nonferrous Alloys and Special-Purpose Materials*, vol. 2, 216–40. ASM Handbook. Materials Park, OH: ASM International.

Young, M., S. Schnepp, F. Casadio, A. Lins, M. Meighan, J. B. Lambert, and D. C. Dunand. 2009. "Matisse to Picasso: A Compositional Study of Modern Bronze Sculptures." *Analytica land Bioanalytical Chemistry* 395 (1): 171–84.

CHAPTER 6 | STAINLESS STEELS

Stainless steels are ferrous alloys with a minimum of approximately 11% chromium. They owe their corrosion resistance to a passive chromium-rich oxide film that forms naturally on the surface. Although extremely thin and invisible, this protective film adheres firmly and is chemically stable under various conditions. Furthermore, the protective oxide film is self-healing, provided oxygen is available in sufficient quantity. Other alloying elements, particularly nickel, molybdenum, and nitrogen, are usually added to further enhance corrosion resistance and improve mechanical properties.

Corrosion-resistant iron-based alloys were first developed at the beginning of the twentieth century. The very first alloys—martensitic, austenitic, and ferritic grades—were based on a composition-driven technology, in which the introduction of new elements like Ni and N and the lowering of C content played key roles in changing the crystalline structure and the microstructure—and consequently the properties (Parr and Hanson 1965, 8–15; Truman 1985; Pieper 2015). Later, owing to new welding and heat-treatment processes, production-technology-driven alloys, such as precipitation-hardening stainless steels, have been introduced.

The development and production of corrosion-resistant iron- and nickel-based alloys permitted new industrial plants of chemical, petrochemical, refining, and other process industries, as well as dairy, food, and pharmaceutical companies. These alloys made it possible to design and fabricate critical equipment, piping, and storage units operable in the widest range of hazardous environments, often at high temperatures and pressures (Cobb 2010, 194–227).

METALLURGICAL BACKGROUND

To be considered "stainless," steel must contain at least 11% chromium in order to develop a thin, adherent, and passive film upon exposure to air or other oxidizing media. Such a film makes the behavior of stainless steels different from that of carbon steel when exposed to the atmosphere. Corrosion products on carbon steel are not compact, and they present cracks. The metal is therefore continually exposed to the environment (fig. 6.1a). In the case of stainless steel, the mixed oxide layer contains chromium, which stabilizes and "closes" the film, rendering it an efficient physical barrier againt corrosion (fig. 6.1b). The surface film can be conveniently thought of as a chromium oxide containing small amounts of other elements in the alloy, which significantly reduces the corrosion rate (fig. 6.2).

The categorization of stainless steels is based upon the nature of their metallurgical structure: the terms used denote the arrangement of atoms that constitute the grains of the alloy. As stated in chapter 1, solid metals and alloys consist of randomly oriented grains that present a well-defined crystal structure, or lattice. In steels, these structures have been given names such as ferrite, martensite, or austenite, or they can be a duplex mixture of two or more of these designations.

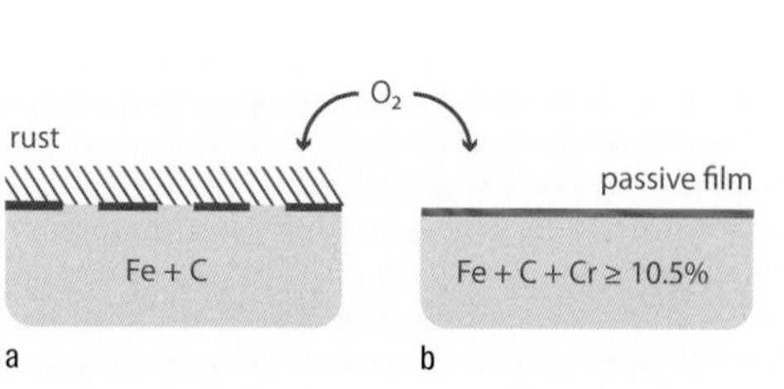

Figure 6.1 | Nonprotective rust layer formed on carbon steel in contact with air humidity or water (a); in contrast, the presence of chromium in stainless steel enables the development of a continuous passive layer (b).

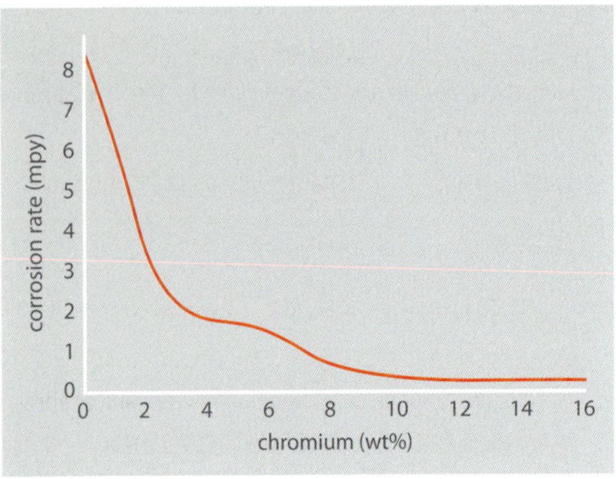

Figure 6.2 | Effect of chromium content on the atmospheric corrosion of low-carbon steel (adapted from AISI Committee of Stainless Steel Producers 1985, fig. 1).

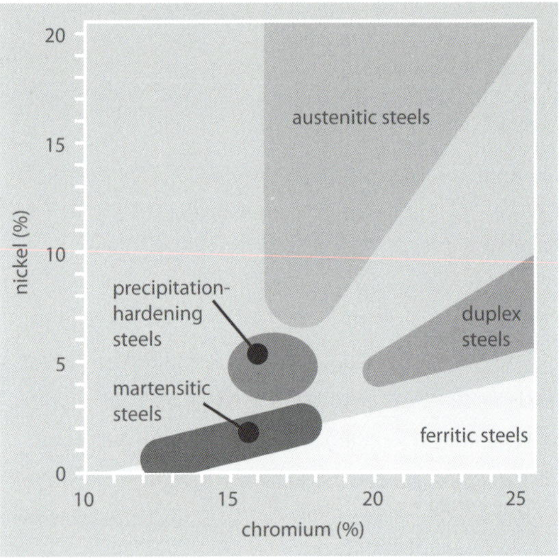

Figure 6.3 | Stainless steel families as a function of nickel and chromium content (adapted from AZO Materials 2017, fig.1).

It is important to understand the differences between these structures because many properties of stainless steels depend upon the crystalline structure.

Ferritic structures are usually obtained for carbon content up to 0.08% (maximum solid solubility in iron at room temperature), but in the case of stainless steels, the presence of chromium in the alloy enlarges the original iron body-centered cubic (BCC) lattice, allowing the accommodation of more carbon atoms in the interstices. Thus, when the chromium content is high enough (11%–27%), the ferritic structure is stabilized, even for carbon contents as high as 0.08%–0.12%. Ferritic stainless steels have properties similar to those of low-carbon steels but present better corrosion resistance because of the presence of chromium.

Martensitic structures are formed when the chromium content is at a lower level (typically 11%–18%) in what is considered the stainless range and the carbon content is higher (typically >0.15%). In this case, the BCC structure is no longer stable, leading to a "stressed" body-centered tetragonal structure (BCT) called martensite. Such high strain energy impedes the motion of dislocations: the material loses ductility and presents the hardest and strongest of the three crystalline forms but becomes the least workable and loses some corrosion resistance as well.

Austenitic stainless steels get their name from austenite, the face-centered cubic (FCC) allotropic form of iron lattice. Because it has larger interstices available, this form can solubilize a greater amount of carbon than the BCC structure can. Although for ordinary steels, austenite exists only at temperatures above 910°C, in the case of stainless steels, the addition of nickel stabilizes it at room temperature. Another way to stabilize austenite is the incorporation of nitrogen, although it can be added only in limited amounts; otherwise, problems such as porosity or formation of chromium nitrides can occur. In such cases, manganese can also be added, and the resulting material is commonly referred to as CrMn stainless steel. In general, austenitic stainless steels are easier to shape and bend, more weldable, and less brittle than ferritic alloys.

In addition to these three classes of stainless steel, other structural possibilities are also commercially exploited. One of them is the family of *duplex steels*, presenting a microstructure that is approximately 50% ferritic and 50% austenitic. This combination gives them a higher strength than either ferritic or austenitic steels. Another is known as *precipitation hardening* (PH) steel, obtained by the further addition of elements such as copper, niobium, and aluminum and the application of a suitable "aging" heat treatment. This process results in the precipitation of very fine particles in the steel matrix, enhancing strength. These steels are particularly useful when a combination of high strength and corrosion resistance is needed.

Depending upon the chemical composition of the steel, the microstructure might be made up of the stable phases austenite or ferrite, a duplex mix of these two, the phase martensite (created when some steels are rapidly quenched from a high temperature), or a structure hardened by precipitated microconstituents (fig. 6.3).

DESIGNATION, PROPERTIES, AND APPLICATIONS

Stainless steels are usually divided into five families according to their microstructure (table 6.1). Each particular grade of stainless steel has its own unique mechanical and physical properties and will usually be produced in accordance with established national or international specifications or standards.

In the past, the different varieties of stainless steels were designated by the American Iron and Steel Institute (AISI) as standard compositions, using a three-digit numbering system. The Unified Numbering System for Metals and Alloys (UNS) is now in use; it allocates one letter ("S") plus five-digit UNS numbers to designate new grades (MatWeb 2017).

In addition to grade, surface finish is an important specification for stainless steel products and is critical in applications in which appearance matters. A smooth surface allows an adequate development of the passive film and provides optimal corrosion resistance. Certain surface finishes also make stainless steel easier to clean, a factor that is clearly important for sanitary applications.

Standard finishes—resulting from passing the steel through rolls or dies—are used in most stainless steel applications. They are categorized as either mill or polished, with reference to surface reflectivity, which varies from dull to mirrorlike in appearance. Mill finishes are the least expensive option. Polished finishes are produced by mechanical abrasion of the surface with a series of abrasives. The smoothest polished finishes are buffed after mechanical polishing to produce a mirrorlike appearance. Sheet finishes are designated by a system of numbers: 1, 2D, and 2B for mill finishes, and 3, 4, 6, 7, and 8 for polished finishes (Davis 1994).

Special finishes include those used when aesthetic appearance is important as well as for particular industrial applications. They include nondirectional scratch patterns, ground circles, swirls, embossed or coined textures, and various coloring and coating methods. Examples of applications and selection criteria for different types of finishes can be found in the literature (SSINA, n.d., 1–16; Zahner 1995, 221–36).

Table 6.1. | General applications and properties of the stainless steel families.

Family	Applications (common alloys)	Hardenability	Corrosion resistance	Magnetic response	Ductility	Weldability
Ferritic	Automotive and appliance trim and flatware (S43000, S44400, S43020)	N	M	Y	M	L
Martensitic	Knife and shear blades, valve and compressor parts (S41000, S43100, S41600)	Quench, temper	M	Y	L	L
Austenitic	Piping and tubing for processing of industrial products, washbasins, cooking utensils (S30200, S30403, S31600)	Cold work	H	N (usually)	VH	VH
Duplex	Chemical, processing, and petroleum industries (S32900)	N	VH	Y	M	H
PH	Golf club heads (S17400)	Aging	M	Y	M	H

Source: Data from AZO Materials 2017.
Note: Y = yes, N = no, L = low, M = medium, H = high, VH = very high.
[a] Corrosion resistance varies significantly between grades within each family—e.g., free machining grades have lower corrosion resistance; grades higher in molybdenum have higher resistance.
[b] Magnetic response: attraction of steel to a magnet; some grades can be attracted to a magnet if cold-worked.

DETERIORATION

As previously stated, the corrosion resistance of stainless steels is due to the presence of a thin oxide film on the surface of the metal. The film's composition varies from alloy to alloy; additionally, it varies with treatments, such as rolling, pickling, and heating. Stabilized by chromium, the film is considered to be continuous, nonporous, insoluble, and self-healing.

When coated by such a film, the steel is said to be passive. However, passivity is not a permanent state; it exists in certain environments or under certain conditions, and it may be lost by slight changes in those conditions. Under circumstances favorable to passivity, stainless steel behaves like a noble metal. General corrosion or uniform attack (in the form of carbon or cast-iron rust) never occurs on stainless steel, except in extremely aggressive conditions. However, as soon as passivity is broken, the alloy reacts locally like ordinary steel (Covert and Tuthill 2000; Sandvik 2017).

A common form of stainless steel corrosion is pitting, in which a highly localized attack occurs at individual sites where the passive layer is discontinuous (fig. 6.4). Once initiated, the corrosion process may proceed very quickly, and it progresses in depth since the rest of the surface is still covered by the passive film. Pitting usually takes place in neutral or acidic solutions primarily containing chlorides (Cl^-). It is activated by temperature and the presence of oxidizing agents.

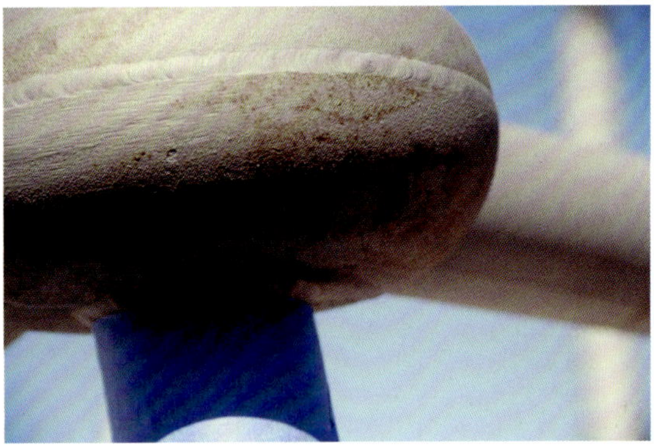

Figure 6.4 | Pitting corrosion on *That Profile* (1999), a stainless steel sculpture by Martin Puryear, Getty Center. Photo: Jane Bassett, 2000.

Crevice or poultice corrosion is another form of localized attack that occurs under stagnant conditions created at the interface of two overlapping materials—the metal surface and another substance, generally one that is not electrically conductive, such as a dirt deposit or gaskets (fig. 6.5). Corrosion usually spreads into the crevice beyond the point of contact, and its progression can be very intense and rapid. The process is activated by the presence of chlorine ions, elevated temperatures, oxygen or easily reducible ions such as ferric ions, and low pH values. Normally, stainless steel should not be painted, because crevice corrosion will result as soon as the paint is damaged.

Intergranular corrosion is observed when the steel has been exposed to a critical temperature range (ca. 450°C–800°C), often because of improper annealing, stress relieving, or heating during forming or welding. At this temperature range, chromium carbides diffuse and precipitate at grain boundaries, creating a galvanic couple with the adjacent chromium-depleted zone. Such a structure is said to be *sensitized* and is more susceptible to attack in some media, particularly acids. When thermal manipulation at the critical range is necessary, sensitization can be avoided by preventing the formation of chromium carbides. This may be achieved by using a low-carbon grade ("L" grade) or by using steel containing titanium or niobium, which preferentially combine with carbon.

Stress-corrosion cracking (SCC) results from the combined action of tensile stresses and a corrosive medium. The process is usually triggered by residual stresses introduced during previous forming and welding operations and not adequately released in the sequence (fig. 6.6). Cracking can be transgranular (across the grains) or intergranular (at grain boundaries) when there is carbide precipitation at those locations.

Galvanic corrosion is not usually a problem for stainless steels, but it can affect the other metals in contact with them. In the passive state, stainless steel behaves like a noble metal, but when the passive film is disrupted, stainless steel can become the anodic metal and corrode in an active manner (see table 2.6). Most galvanic corrosion problems can be avoided by proper design or electrical insulation. When contact is unavoidable, bolts and rivets intended to connect a stainless steel sheet or plate should be made of a more noble alloy.

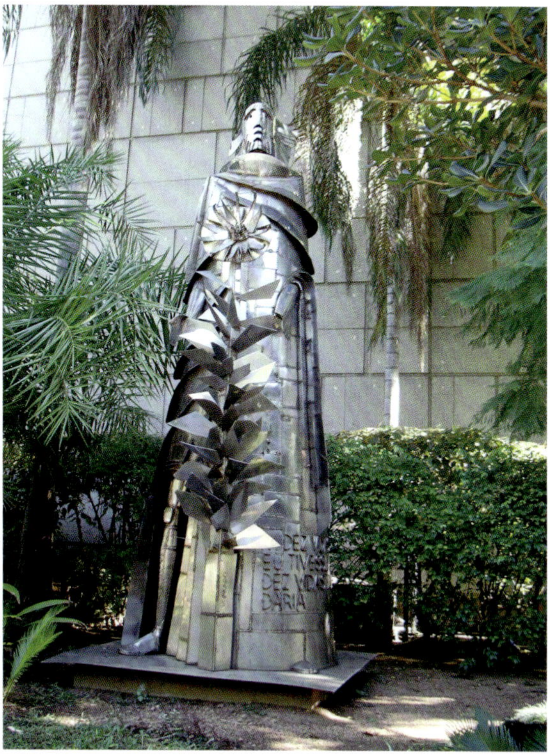

a

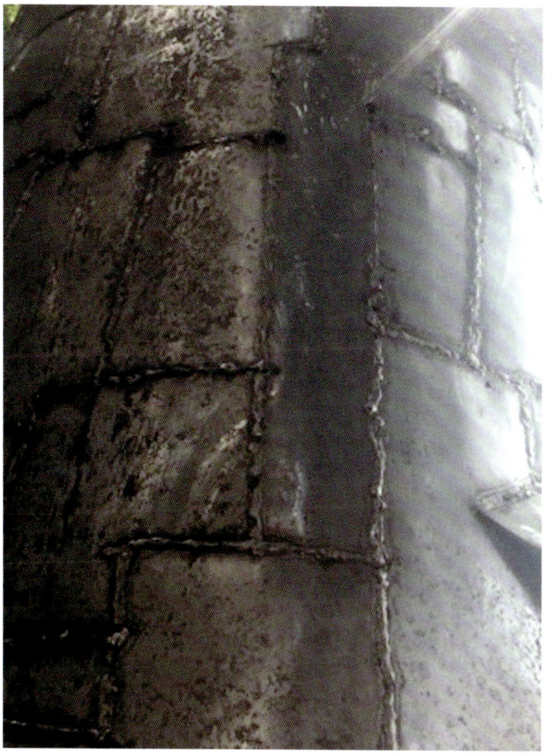

b

Figure 6.5 | Stainless steel sculpture *Tiradentes* by Vasco Prado (1976), Porto Alegre, Brazil (a); region presenting different surface conditions: On the right side, the surface is clean and shiny because it is regularly washed by rain, while on the left side, it is partially covered by deposits from pollution and vegetation. After removal of the deposits, the underlying surface will likely show some kind of localized corrosion. Photos: Virginia Costa.

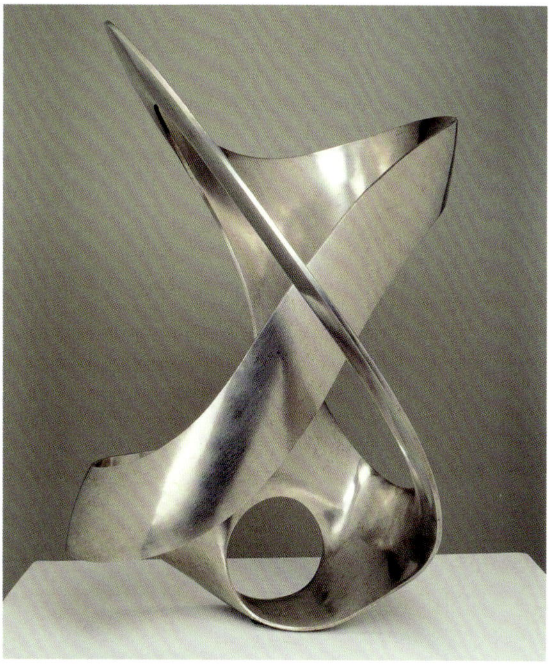

a

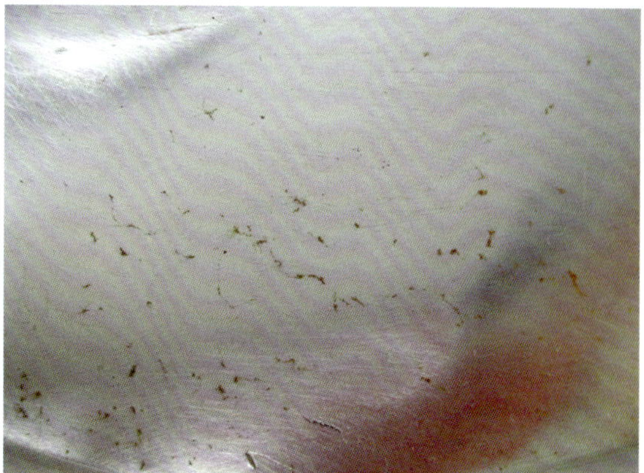

b

Figure 6.6 | Stainless steel sculpture *Unidade Tripartida* by Max Bill (1948–49), Museu de Arte Contemporânea da Universidade de São Paulo (a); detail showing a network of oxidized fine cracks, very likely due to residual stresses (b). Photos: Virginia Costa.

CHARACTERIZATION

Stainless steel can be easily differentiated from carbon steel because it is rarely painted, and its surface does not show rust stains unless it has been contaminated with ferrous particles. Moreover, austenitic grades are not attracted by a permanent magnet, showing only a weak attraction if cold-worked (severely bent or rolled).

During inspection for conservation condition, particular attention must be paid to possible localized corrosion. This type of deterioration is difficult to assess because of the small size of its features, which are especially hard to see when covered by deposits or established in crevices (see fig. 6.5b).

CONSERVATION-RELATED LITERATURE

Although the presence of stainless steels in quite diverse cultural heritage collections (art, technology, buildings, equipment) is large and expanding rapidly, publications detailing techniques used or decisions made regarding surface interventions are rare (Nagy and Zahner 2013; Considine et al. 2010; Hofmeister, Decker, and Bührer 2014). Stainless steel has often been reported as a replacement structural part for buildings (Baptiste 1994; Nacheman 2006; Euro Inox 2007), sculpture (Russel 2004), ships (Kulchar 2013), and even a human skeleton (Caldararo, Antonetti, and Hirschbein 2014). Some research has been done to investigate its performance in contact with various materials (Glavan 2004; Costa and Texier 2010).

REFERENCES AND BIBLIOGRAPHY

AISI [American Iron and Steel Institute] Committee of Stainless Steel Producers. 1985. *Stainless Steels for Machining*. Washington, DC: American Iron and Steel Institute.

American Iron and Steel Institute. 1988. *Cleaning and Descaling Stainless Steels*. Designers' Handbook no. 9001. Washington, DC: American Iron and Steel Institute. *https://www.nickelinstitute.org/~/Media/Files/TechnicalLiterature/CleaningandDescalingStainlessSteel_9001_.pdf*.

ASTM International. 2013. *ASTM A380M. Standard Recommended Practice for Cleaning and Descaling Stainless Steel Parts, Equipment and Systems*. West Conshohocken, PA: ASTM International.

AZO Materials. 2017. "Stainless Steels: Introduction to the Grades and Families." *https://www.azom.com/article.aspx?ArticleID=470*.

Baptiste, H. 1994. "Restauration du Pavillon suisse de Le Corbusier, Cité internationale universitaire, Paris." *Monumental*, no. 5 (March): 71–79.

British Stainless Steel Association [BSSA]. 2001. *Care and Maintenance of Stainless Steel*. SSAS information sheet no. 7.20. Sheffield: British Stainless Steel Association.

Buijs, N. W. 2006. "Dry Ice Blasting of Stainless Steel." *Stainless Steel World*, April: 31–35. *http://gilsen.com/sites/default/files/files/stainlesssteel.pdf*.

Caldararo, N., C. Antonetti, and J. Hirschbein. 2014. *The Restoration of a Human Skeleton: A Scientific System*. SPNHC leaflet no. 6. New York: Society for the Preservation of Natural History Collections.

Clapper Morris, J., and D. Slaton. 2015. "Modern Metals: Finishes, Investigation, and Conservation Considerations." Special Issue on Modern Metal Finishes, *APT Bulletin* 46 (1): 29–38.

Cobb, H. M. 2010. *The History of Stainless Steel*. Materials Park, OH: ASM International.

Considine, B., J. Wolfe, K. Posner, and M. Bouchard. 2010. *Conserving Outdoor Sculpture: The Stark Collection at the Getty Center*. Los Angeles: Getty Conservation Institute.

Costa, V., and A. Texier. 2010. "Restoration of Cultural Heritage: Evaluation of the Compatibility between Metals and Sealing Products." *Journal of Solid State Electrochemistry* 14:403–5.

Covert, R. A., and A. H. Tuthill. 2000. "Stainless Steels: An Introduction to their Metallurgy and Corrosion Resistance." *Dairy, Food and Environmental Sanitation* 20 (7): 506–17.

Davis, J. R. 1994. "Surface Engineering of Stainless Steels." In *Surface Engineering*, edited by ASM Handbook Committee, 741–61. Vol. 5 of *ASM Handbook*. Materials Park, OH: ASM International.

Euro Inox. 2007. *Renovation of the Atomium in Brussels*. Brussels: European Stainless Steel Development Association.

Glavan, J. 2004. "An Evaluation of Mechanical Pinning Treatments for the Repair of Marble at the Second Bank of the United States." MSc diss., University of Pennsylvania.

Hofmeister, S., P. Decker, and A. Bührer. 2014. "Restaurierung in luftiger Höhe: bildsäulen-dreiergruppe: Eine 32 Meter hohe Außenskulptur von Max Bill." *Restauro* 119 (1): 50–57.

Kulchar, D. 2013. "Super Duplex to Keep the Vasa Safe." *MolyReview*, no. 1. *http://www.imoa.info/download_files/molyreview/excerpts/13-1/Vasa.pdf*.

MatWeb. 2017. "Metal Alloy UNS Number Search." *http://www.matweb.com/search/SearchUNS.aspx*.

NAAMM and NOMMA [National Association of Architectural Metal Manufacturers and National Ornamental and Miscellaneous Metals Association]. 2006. "Finishes for Stainless Steel." In *Metal*

Finishes Manual for Architectural and Metal Products, 3-14 to 3-18. Chicago: National Association of Architectural Metal Manufacturers. *http://www.nomma-northeast.org/wp-content/uploads/2010/11/NAAMM-NOMMA-Finishes-Manual.pdf.*

Nacheman, R. J. 2006. "The Empire State Building: Façade Evaluation and Repair of an Engineering Landmark." *Structure Magazine*, January: 39–43. *http://www.structuremag.org/wp-content/uploads/2014/09/Empire-State-Building1.pdf.*

Nagy, E., and W. Zahner. 2013. "Remediation of Impact Damage on Modern and Contemporary Sheet Metal Sculpture: Conservation Treatment of Donald Judd's 'Stack.'" In *Metal 2013: Edinburgh, Scotland: Interim Meeting of the ICOM-CC Metals Working Group, 16–20 September 2013, Conference Proceedings*, edited by E. Hyslop, V. Gonzalez, L. Troalen, and L. Wilson, 283–92. Edinburgh: Historic Scotland.

Parr, J. G., and A. Hanson. 1965. *An Introduction to Stainless Steel*. Materials Park, OH: ASM International.

Pieper, R. 2015. "The 'White Metals' of Early-Twentieth-Century American Architecture." Special Issue on Modern Metal Finishes, *APT Bulletin* 46 (1): 23–28.

Russel, K. 2004. "The Case of the Lamed Lady Liberty." *Design News* 59 (1): 124.

Sandvik. 2017. "Corrosion Tables." *http://smt.sandvik.com/en/materials-center/corrosion-tables/.*

Sedricks, A. J. 1979. *Corrosion of Stainless Steels*. New York: John Wiley and Sons.

SSINA [Specialty Steel Industry of North America]. n.d. *Special Finishes for Stainless Steel*. Designer Handbook. Washington, DC: Specialty Steel Industry of North America. *http://www.ssina.com/download_a_file/special_finishes.pdf.*

Truman, J. 1985. "The Initiation and Growth of High Alloy (Stainless) Steel Production." *Historical Metallurgy* 19 (1): 116–25.

Zahner, L. W. 1995. *Architectural Metals: A Guide to Selection, Specification and Performance*. New York: John Wiley and Sons.

CHAPTER 7 | TITANIUM ALLOYS

Although discovered in the eighteenth century, titanium was purified only in the early twentieth century and was first used as a structural metal in 1952. It is strong and has a specific density about halfway between aluminum and steel. Because of their high strength-to-weight ratio and good corrosion resistance, titanium alloys were first employed in the aerospace and chemical industries, where the combination of these unique properties justified the high cost. The decrease of titanium-alloy in price associated with their increasing availability has widened their use over competing materials such as aluminum alloys and steels. Titanium alloys are available as castings, powder metallurgy (PM) products, and all wrought forms, such as plates, sheets, tubes, forgings, bars, and wires. Current applications include industrial equipment, process instrumentation, medical implants, sports goods, jewelry, sculpture (Chiles 1987), and architecture (fig. 7.1) (Peters and Leyens 2003; Cacace, Nikaki, and Stefanidou 2012).

METALLURGICAL BACKGROUND

Titanium exists in two crystallographic forms, known as alpha and beta, which have important implications regarding properties. The alpha phase exists at room temperature and has a

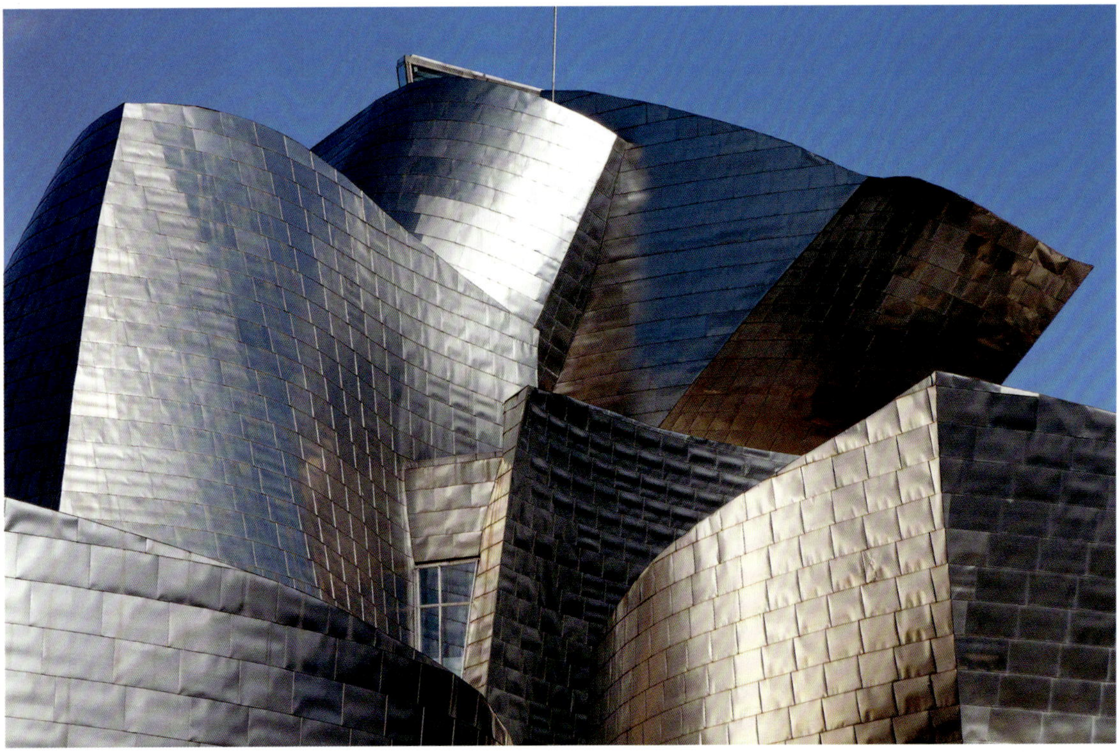

Figure 7.1 | Titanium cladding of the Guggenheim Museum, Bilbao, giving the building a unique visual effect.

hexagonal close-packed (HCP) structure. The beta phase has a body-centered cubic (BCC) structure, is more ductile owing to the larger number of slip planes, and is found at high temperatures only, above about 880°C. The critical temperature of alpha-to-beta phase transition can, however, be modified by the addition of alloying elements. Alpha stabilizers are elements that raise the critical temperature—as, for instance, aluminum, oxygen, and nitrogen—stabilizing the alpha crystal structure. Conversely, other elements, such as vanadium, molybdenum, chromium, and iron, lower the transition temperature and are called beta stabilizers. Manipulating the relative amount of these two allotropic phases through alloy additions and thermomechanical processing is commonly done to provide a combination of useful properties (Donachie 2000, 2–3).

Titanium alloys are classified as alpha, near alpha, alpha-beta, and beta according to the nature and content of alloying elements and phases (fig. 7.2). *Alpha alloys* are single-phase alloys. They are solid solutions strengthened by the addition of alpha stabilizers or neutral alloying elements. They have good stability but are not amenable to heat treatment for property modifications. *Near alpha alloys* are primarily alpha alloys containing a small amount of beta phase in the final microstructure, owing to additions of 1%–2% of beta stabilizers to improve strength and workability. *Alpha-beta alloys* contain larger amounts of beta stabilizers (4%–6%). They can be heat-treated to develop a variety of microstructures and mechanical properties. *Beta alloys* contain very large additions of beta stabilizers (30%) to make the beta phase stable at room temperature. However, they present higher density and poor ductility. Beta alloys are usually metastable, which means that even slight thermic or mechanical stress may induce partial transformation to alpha or other phase (Donachie 2002).

The final strength of titanium alloys depends on chemical composition as well as on the deformation processes used to shape them. The grain size and shape and the arrangement of grain boundaries have a highly significant impact on mechanical properties.

Titanium surfaces are very reactive. They have an extremely high affinity for oxygen, forming spontaneously and instantly a highly adherent, continuous, and stable oxide film when exposed to air and/or moisture. In most aqueous environments, the oxide is close to stoichiometric TiO_2 but may consist of mixtures of other titanium oxides (Davis 2001, 417–31).

DESIGNATION, PROPERTIES, AND APPLICATIONS

Currently available commercial titanium and titanium alloys are unalloyed grades (commercially pure) and alloyed grades (alpha and near alpha alloys, alpha-beta alloys, and beta alloys). Selection of an unalloyed grade or higher-strength alloy depends on desired mechanical properties, service requirements, and processing considerations, among other factors.

Unalloyed titanium is usually chosen for its excellent corrosion resistance, especially in applications where high strength is not required. The primary differences between unalloyed grades lie in oxygen and iron content. Grades of higher purity (lower interstitial content) present lower strength and hardness but have greater formability.

The various titanium grades as defined by ASTM are numbered from 1 upward. Table 7.1 presents some commonly used alloys. Grades 1–4 are unalloyed titanium (commercially pure) and have an ascending content of impurity. Grade 1 is the most ductile, while grade 4 has the best mechanical properties. Grade 5 is assigned to one of the most commonly used alloys, belonging to the alpha-beta family, Ti-6Al-4V (UNS R56400). These alloys can be strengthened by heat treatment or by thermomechanical

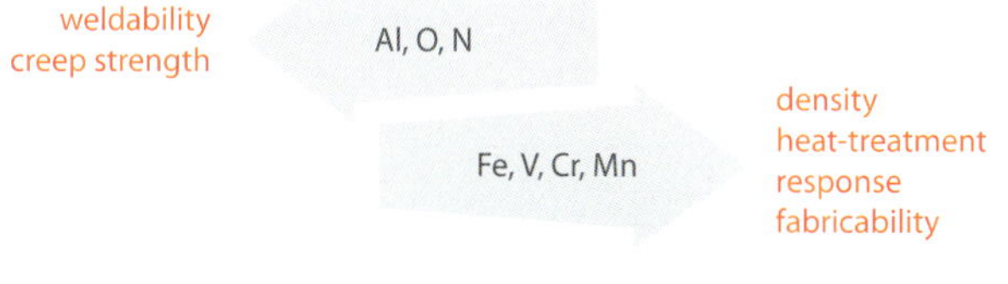

Figure 7.2 | Effect of alloying elements on various properties (adapted from Donachie 2000, fig. 1.4).

Table 7.1. | Titanium alloys in common use.

Titanium alloy	UNS designation (ASTM)	Composition	Properties	Applications
Unalloyed (alpha)	R50250–R50700 (Grades 1–4)	Small additions of O and Fe	Corrosion resistance, formability, and weldability	Aircraft, marine, and chemical industries
Alloyed (alpha-beta)	R56400 (Grade 5)	6% Al + 4% V	Good creep resistance, excellent fatigue strength, and fair weldability	Aircraft turbine parts, structural components, prosthetic implants (most widely used titanium alloy)

Source: Data from Lampman 1990.

processing to produce a variety of microstructures offering a wide range of properties, with the usual trade-off between strength and ductility.

As regards surface finish, titanium anodization is a very popular technique. A large variety of interference colors may be obtained by anodizing at different voltages to obtain decorative effects on jewelry and other items, as well as for product identification in medical devices (see fig. 11.4). The aerospace, military, and machine tooling industries make use of titanium anodizing to improve antigalling, wear resistance, and corrosion resistance (SAE 2018; Diamanti, Del Curto, and Pedeferri 2006; O'Hana et al. 2008).

DETERIORATION

Titanium is a reactive material, and its corrosion resistance depends on the presence of the protective oxide film, which is thermodynamically stable in a wide range of conditions. As a result, titanium alloys are expected to resist all natural waters (including seawater) and salt solutions containing common anions—that is, chlorides, sulfates, silicates, phosphates, nitrates, carbonates, and so on—of pH 3–12. However, under moderate to strong reducing conditions in acidic media (HCl, H_2SO_4, or H_3PO_4 solutions), oxide film breakdown occurs, resulting in general corrosion, and especially severe attack occurs in hydrofluoric acid solutions, even in relatively low concentrations. Small additions of alloying constituents (<2%–3%) have little effect on the basic corrosion resistance of Ti in normally passive environments (TMC 2007).

Regarding particular forms of corrosion, titanium alloys are prone to crevice corrosion where tight cavities exist, especially in aqueous halide solutions. Pitting is observed when iron or steel particles become embedded in a titanium surface, breaking the oxide film. Localized attack initiated by this mechanism creates a very characteristic circular pit morphology. Regarding galvanic corrosion, under normally passive conditions, titanium presents a quite noble corrosion potential, similar to stainless steel or nickel-based alloys in their respective passive states. In most environments, Ti is the cathodic member of any galvanic couple and can accelerate the corrosion of the other member. Titanium alloys with oxygen and aluminum can be especially susceptible to stress-corrosion cracking (SCC) if residual tensile stresses persist after cold work fabrication or under externally applied load (Schulz 2005).

CHARACTERIZATION

Although titanium alloys can present many colors owing to various surface finishes, the bare metal is a shiny dark silver. Compared to freshly scratched aluminum, scratches in titanium alloys should appear much darker. They are nonmagnetic and have a specific density of 4.5 g/cm^3—halfway between aluminum and steel, as stated above. Titanium alloys, being expensive, are not used in the production of low-quality items.

CONSERVATION-RELATED LITERATURE

The very few references to titanium alloys in the conservation literature are related to their use as structural reinforcement for the restoration of the marble monuments on the Acropolis (Ioannidou 2007; Zambas, Ioannidou, and Papanikolaou 1986; Kourkoulis and Ganniari-Papageorgiou 2010; Skoulikidis and Vassiliou 2000).

REFERENCES AND BIBLIOGRAPHY

ASTM International. 2011. *ASTM B600-11. Standard Guide for Descaling and Cleaning Titanium and Titanium Alloy Surfaces*. West Conshohocken, PA: ASTM International.

Cacace, K., M. Nikaki, and A. Stefanidou. 2012. "Guggenheim Museum Bilbao: An Evaluation of the Cladding Materials." *http://www.academia.edu/29271196/GUGGENHEIM_MUSEUM_BILBAO_AN_EVALUATION_OF_THE_CLADDING_MATERIALS*.

Chiles, J. R. 1987. "Titanium the Magnificent." *Smithsonian* 18 (2): 87.

Davis, J. R. 2001. *Alloying: Understanding the Basics*. Materials Park, OH: ASM International.

Diamanti, M. V., B. Del Curto, and M. Pedeferri. 2006. "Interference Colors of Thin Oxide Layers on Titanium." *Color Research and Application* 33 (3): 221–28.

Donachie, M. J. 2000. *Titanium: A Technical Guide*. Materials Park, OH: ASM International.

——. 2002. "Selection of Titanium Alloys for Design." In *Handbook of Materials Selection*, edited by M. Kutz, 201–34. New York: John Wiley and Sons.

Ioannidou, M. 2007. "Principles and Methodology of Intervention for Structural Restoration." In *Proceedings of the XXI International CIPA Symposium, 1–6 October, Athens, Greece. http://www.isprs.org/proceedings/XXXVI/5-C53/*.

Kourkoulis, S. K., and E. Ganniari-Papageorgiou. 2010. "Restoring Fragmented Marble Epistyles: Some Critical Points." *Journal of Cultural Heritage* 11:420–29.

Lampman, S. 1990. "Wrought Titanium and Titanium Alloys." In *Properties and Selection: Nonferrous Alloys and Special-Purpose Materials*, 592–633. Vol. 2 of *ASM Handbook*. Materials Park, OH: ASM International.

O'Hana, S., A. J. Pinkerton, A. W. Gale, and L. Li. 2008. "Laser Surface Colouring of Titanium for Contemporary Jewelry." *Surface Engineering* 24 (2): 147–53.

Peters, M., and C. Leyens. 2003. "Non-Aerospace Applications of Titanium and Titanium Alloys: Architecture." In *Titanium and Titanium Alloys*, edited by C. Leyens and M. Peters, 405. Weinheim, Germany: Wiley-VCH Verlag.

SAE [Society of Automotive Engineers]. 2018. *AMS2488D. Anodic Treatment: Titanium and Titanium Alloys Solution pH 13 or Higher*. Warrendale, PA: SAE International. *https://www.sae.org/standards/content/ams2488d/*.

Schulz, R. W. 2005. "Titanium." In *Corrosion Tests and Standards Manual*, edited by R. Baboian, 598–612. Baltimore: ASTM International.

Skoulikidis, T., and P. Vassiliou. 2000. "Damages in Monuments Produced by the Corrosion of Metallic Junctions: The Acropolis Case." In *Proceedings of the 9th International Congress on Deterioration and Conservation of Stone, Venice, June 19–24*, edited by V. Fassina, 155–63. Amsterdam: Elsevier Science.

TMC [Titanium Metals Corporation]. 2007. *TIMET®: Corrosion Resistance of Titanium*. Denver: Titanium Metals Corporation. *http://www.timet.com/assets/local/documents/technicalmanuals/corrosion.pdf*.

Zambas, C., M. Ioannidou, and A. Papanikolaou. 1986. "The Use of Titanium Reinforcement for the Restoration of Marble Architectural Members of the Acropolis Monuments." *Studies in Conservation* 31:138–41.

CHAPTER 8 | WEATHERING STEELS

The term weathering steel—structural steel with improved atmospheric corrosion resistance—describes a class of steels that contain small amounts of alloying elements such as Cu, Cr, P, Ni, Si, and Mn. Their presence enhances the steels' corrosion resistance in less aggressive atmospheres by the formation of a compact and well-adhering corrosion-product layer (patina), permitting them to be used outdoors without paint. The patina imparts a pleasing dark surface to weathering steels. Applications include civil structures such as bridges and other load-bearing structures, road installations, electricity posts, utility towers, guide rails, ornamental sculptures, and facades.

Weathering steels are descendants of the copper-bearing steels that came into use early in the twentieth century (Morcillo et al. 2013). It was reported that when exposed to the atmosphere, some copper-containing irons and steels corroded more slowly than others. Additional studies showed that in addition to copper, small amounts of other alloying elements provided greater atmospheric corrosion resistance and enhanced strength. In 1933 United States Steel introduced COR-TEN (a trademark formed from "corrosion-tensile"). Because of its exceptional strength and corrosion resistance, it became possible to use less material for the same strength specification as well as reduce costs by the elimination of a protective coating, which requires frequent maintenance. In the early 1960s, COR-TEN was promoted in the architectural and civil engineering literature as an alloy that combined structural and aesthetic utility, and its adoption led to sculptural applications within a few years (Scott 1993; Morcillo et al. 2014).

Today weathering steels are one of the many categories of high-strength low-alloy (HSLA) steels, which include standard and proprictary grades designed to provide specific desirable combinations of properties such as strength, toughness, formability, weldability, and atmospheric corrosion resistance. They are not considered alloy steels in the normal sense because they are designed to achieve specific properties rather than to conform to a chemical composition, which may vary for different product thicknesses to meet property requirements.

METALLURGICAL BACKGROUND

In most environments, ferrous alloys have a tendency to rust at a rate that depends on the access of oxygen, moisture, and atmospheric contaminants to the metal surface. On most ordinary structural steels, rust layers are porous, and after a certain period of time, they detach from the surface, reexposing the metal, which therefore continues to corrode.

With weathering steel, the rusting process is initiated in the same way, but the specific alloying elements in the steel produce a more stable rust layer; it is much less porous and adheres to the base metal. Such a patina must develop under conditions of alternating wetting and drying in order to produce a protective barrier. The result is a much lower corrosion rate than would be found on ordinary structural steel (fig. 8.1).

Since the corrosion performance of weathering steel relies on the stability of the patina, parameters that may favor its formation—steel composition and environmental conditions—play an important role and must be considered in design as well as maintenance.

The corrosion-retarding effect of the protective layer is due to the particular distribution and concentration of the alloying elements, such as copper, chromium, nickel, and manganese. These elements, which are present in very small amounts, catalyze the formation of particularly stable iron corrosion products and do not form mixed-metal oxides, as is the case with stainless steels (Wurth 1991; Angelini et al. 2011). The rust formed on the weathering steel is basically a bilayer.

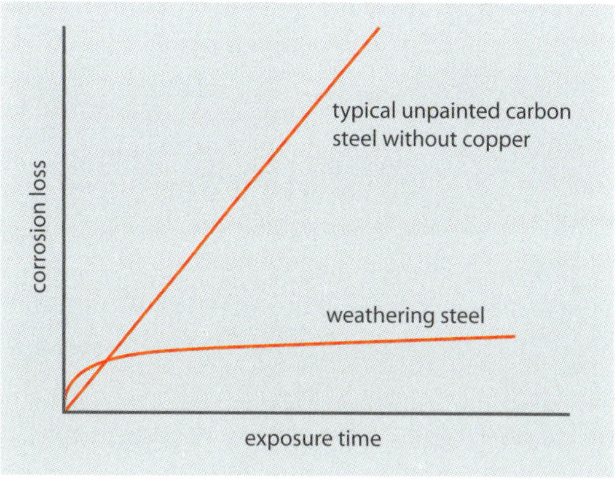

Figure 8.1 | Comparison of corrosion rates for weathering steel and carbon steel, both unprotected (adapted from ArcelorMittal 2017a).

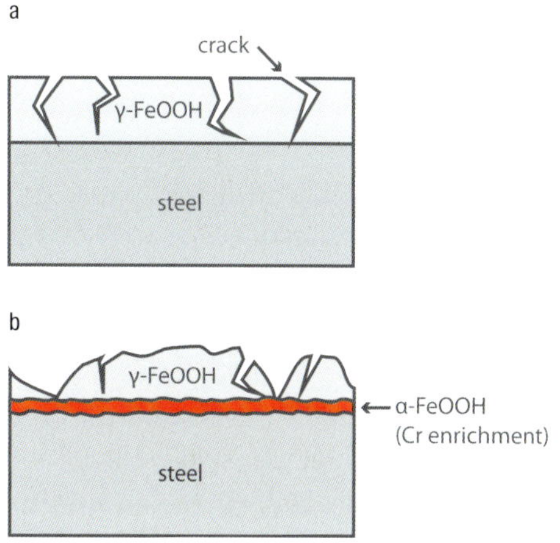

Figure 8.2 | The development of a stable and protective rust layer on weathering steel in industrial regions: Initial rusting (a); and after the passage of decades, the final, stable stage (b) (adapted from Yamashita et al. 1994, fig. 11).

The outer layer is loose and often flaky, mainly composed of gamma-FeOOH (lepidocrocite). The inner layer consists of fine, tightly adherent amorphous and densely packed nanoparticles of alpha-FeOOH (goethite). It has been proposed that the lepidocrocite forms first, before being progressively transformed into the more chemically stable goethite during the long-period atmospheric corrosion (Yamashita et al. 1994, Yamashita et al. 2004; Asami and Kikuchi 2003) (fig. 8.2).

Research on the nature of this protective layer attributes its characteristics to the presence of the alloying elements: Cu and P (Misawa et al. 1971, Misawa et al. 1974; Aramendia et al. 2013) and Cr (Wurth 1991; Yamashita et al. 1994, Yamashita et al. 2004; Qian et al. 2013). Moreover, recent studies on advanced weathering steels containing nickel conclude that its presence prevents Cl ions from reaching the steel surface and thereby increases the corrosion resistance (Kimura et al. 2005).

The initial surface condition is another important factor contributing to the formation of a protective patina. Steels are generally delivered with a surface covered by mill scale, a thin, adherent oxide coating that forms during heat treatment, hot rolling, and forging processes. This layer offers very little long-term protection to atmospherically exposed steel and must be removed prior to use, in order to expose the bare steel to the local environment so as to permit formation and growth of an adherent rust layer (Castro et al. 2011). Similarly, artificial patinas applied on the surface before exposure may offer poor protection against alloy dissolution (Chiavari et al. 2012; Ramirez-Barat et al. 2016).

Finally, the ability of weathering steels to fully develop their anticorrosive action depends on climate and exposure conditions. Structures should be free of interstices, crevices, cavities, and other places where water can stagnate, allowing corrosion to progress without the formation of a protective patina. It is now well accepted that wet-dry cycling is necessary for the formation of a dense and adherent rust layer. This involves rainwater washing the steel surface well, accumulated moisture draining easily, and a rapid drying action (absence of very long periods of wetness). The development of a protective layer is gradual, requiring several years before steady-state conditions are obtained (fig. 8.3) (Leygraf and Graedel 2000).

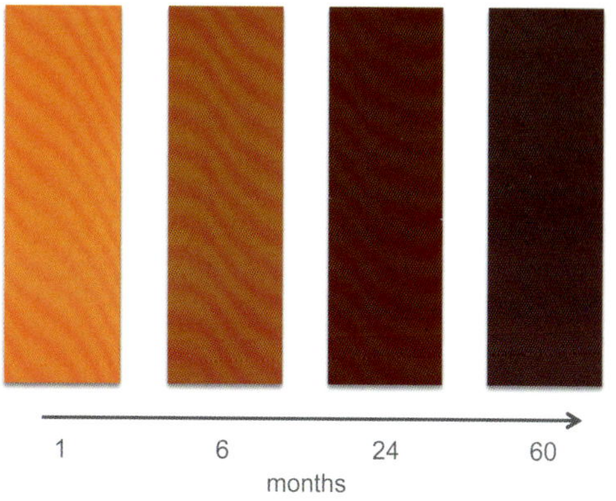

Figure 8.3 | Gradual color change of protective rust in general application environments. After two years, the coloration shows no clear change (adapted from Nippon Steel and Sumimoto Metal Corp. 2017).

DESIGNATION, PROPERTIES, AND APPLICATIONS

The term weathering steel does not correspond to a defined range of chemical composition but, rather, encompasses steels with a variable range of composition, provided that they satisfy minimum levels of yield strength and corrosion resistance. They are described in different ASTM specifications, depending on the type of product (ASTM International 2013, 2015a). The most prevalent grades are ASTM A242, loosely stated as Corten A, architectural grade (UNS K11510), and ASTM A588 Grade A, equivalent to Corten B, structural grade (UNS K11430).

The chemical composition varies for different product types (plates, sheets, coils, tubes, or bars) to meet property requirements but shall conform to limits imposed by the corresponding specification. Choice and use of alloying elements are determined by the manufacturer, provided that the elements combined with carbon, manganese, phosphorus, sulfur, and copper within the limits specified, give the prescribed mechanical properties and corrosion resistance. Elements commonly added include Cr, Ni, Si, V, Ti, and Zr and must be reported in the heat analysis sheet to identify the type of steel. The atmospheric corrosion resistance index is estimated on the basis of chemical composition and from actual exposure tests, as described in ASTM guide G101–15 (ASTM International 2015b).

As weathering steels became more popular over the years, different manufacturers began to develop their own atmospheric-corrosion-resistant weathering steels. The content limit in alloying elements has evolved, and new compositions have been developed to improve mechanical properties and for use in more aggressive environments.

DETERIORATION

The structural integrity of the rust layer plays an important role in the overall corrosion process of weathering steels. If it is porous or cracked, corrosion will continue, penetrating the steel. When a weathering steel is subjected to long periods of wetness—staying permanently wet or covered with accumulated products that concentrate moisture—the wet-dry cycling required for the ongoing formation of a protective layer is prevented (Decker, Brüggerhoff, and Eggert 2008). For instance, this is the case upon long-term immersion in freshwater or seawater and burial in soil with varying moisture levels. The lack of drying cycles inhibits the formation of the protective oxide film, resulting in accelerated deterioration (fig. 8.4).

Moreover, the above requirements for developing the protective film must be considered locally because the corrosion rates may vary greatly on different parts of the same object. The surface exposed to the ground is generally more corroded than that exposed to the air, and in the Northern Hemisphere, the surface exposed to the north presents a higher corrosion rate than the surface exposed to the south.

Disturbance of the rust layer by some pollutants can eliminate its protective capacity. For instance, a high level of chloride in the environment generally causes the formation of akaganeite instead of goethite, lowering corrosion resistance

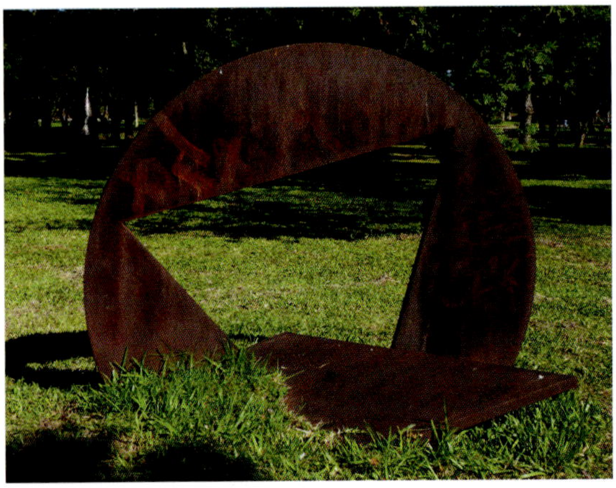

a

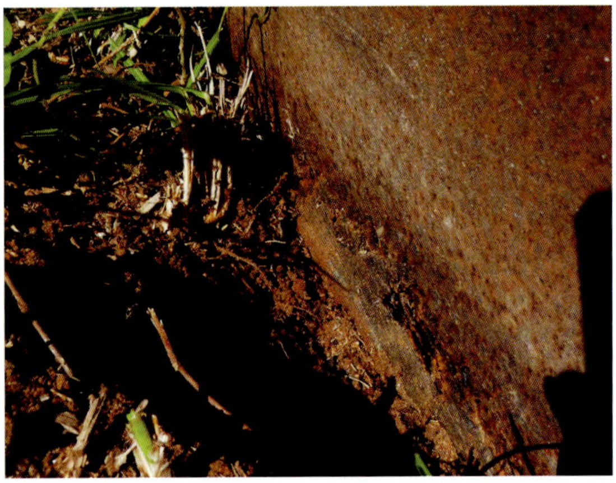

b

Figure 8.4 | Sculpture *Sem titulo* by Amílcar de Castro (1994), Porto Alegre, Brazil (a); detail of the buried part, showing exfoliation (b). Photos: Virginia Costa.

and leading to exfoliation. In areas of severe air pollution with sulfates, surface deposits create localized areas of high acidity that may dissolve the protective oxide. The effect of calcium oxalates on the conservation of weathering steels has also been investigated (Aramendia et al. 2014a).

Finally, as previously mentioned, the preparation of the surface before exposure seems to play an important role in the atmospheric corrosion behavior of different weathering steel compositions (Castro et al. 2011).

Regarding maintenance, periodic visual examination is indispensable to confirm the stability of the patina layer. Particular attention must be paid to the presence of debris lying on the structure or to the proximity of vegetation, because parts in such conditions will remain permanently wet. It is important to check for surfaces contaminated by retained pollution, graffiti, or other material, as well as to look for evidence of stagnant water at any place on the structure. Removing the protective layer locally may be necessary to determine the extent of pitting and to measure the section loss.

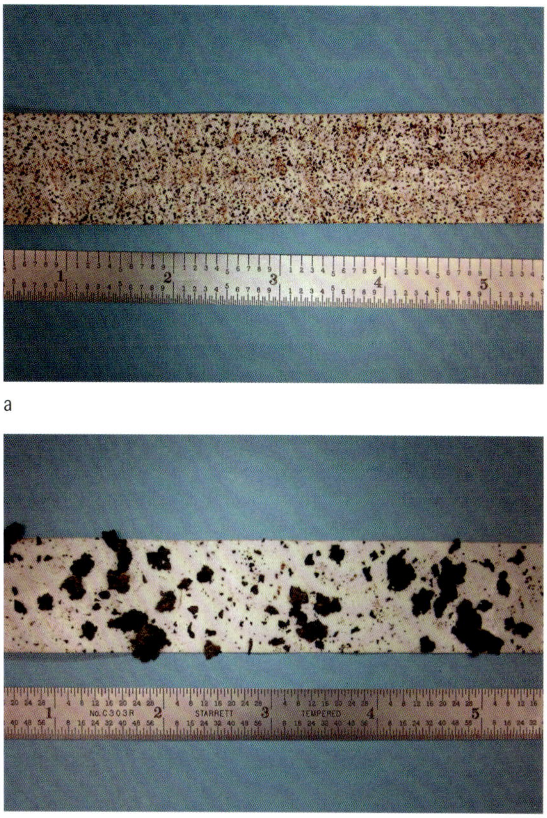

a

b

Figure 8.5. | Examples of tape tests for two patina conditions: Fine reddish rust spots interspersed with dark brown patina flakes (Category 5) (a); and medium (5–25 mm) flakes over most areas easily removed with tape (Category 2) (b) (reproduced from WJE 2013, figs. 23, 26).

CHARACTERIZATION

Weathering steel is usually employed without a coating because the rust layer offers an efficient protection if it is correctly prepared and conserved. The surface color depends on several factors, including steel composition (the higher the content of alloying elements, the darker the color), environmental conditions (unpolluted atmosphere leads to lighter color), and time of exposure. The yellowish appearance in the initial stage gradually changes to an orange-brown color and eventually to a stable dark brown after a few years of exposure. After that point, there is no change, except possibly to a deeper dark brown (see fig 8.3).

Assessment methods for evaluating patina stability include visual inspection, thickness and resistance measurements, and electrochemical techniques. In general, the surface condition of the oxide is a good indicator of performance: an adherent and fine-grained oxide indicates that corrosion is progressing as expected, while coarse, flaking layers of oxide suggest excessive corrosion. Inspection guidelines are available that highlight the structural features to be inspected and illustrate the difference between a protective oxide and excessive rust. Rating systems have been developed that consider either just the rust aspect in a five-step ranking (Kihira et al. 2009) or a set of complementary characteristics—such as patina color, uniformity, adherence, and rust flake size—to arrive at a patina rating index (WJE 2013).

Tape testing, where adhesive tape is applied to a corroded surface in order to assess the material that adheres to it, has been used as a simple method to comparatively assess different parameters (adherence, rust flake size, etc.) within the same structure (WJE 2013). Depending on patina performance, an index is attributed; it provides a consistent quantitative description of the steel's overall performance and facilitates comparison of trends in steel behavior across multiple parts or artifacts or over multiple inspection periods. Moreover, photographing the tape specimens at each inspection is very useful for determining whether the patina condition is deteriorating or stabilizing (fig. 8.5).

Another way to assess the stability of a weathering steel is by monitoring the plate thickness at different locations.

However, this is impossible using mechanical means because the rust layer affects the result. Portable ultrasonic equipment is available to measure residual thickness without removing the rust layer (Sheikh and Bligh 2013). Thickness measurements have been correlated with rust film resistance measurements in order to propose a classification of the protective properties of rust layers (Kihira, Ito, and Murata 1989; Kihira 2007). By these methods, it is possible to establish a diagram defining three main areas: an area where the surface can be considered covered with a protective oxide, an area where the oxide is in an initial developing stage, and an area where the oxide is anomalous and not protective (fig. 8.6) (Deflorian and Rossi 2002).

The protective characteristics of rust layers have also been investigated using electrochemical techniques. The value of the open circuit potential (OCP) is a first indication: when the rust layer is stable, the OCP value is more positive than when the rust is not protective (Pourbaix and Miranda 1976; Townsend, Simpson, and Johnson 1994). Polarization curves have been performed on samples taken from structural components, and they clearly show differences between well-conserved and poorly conserved parts: those performing well during service life presented a current density that was much smaller than the others (Deflorian and Rossi 2002). Electrochemical impedance spectroscopy (EIS) has been performed on samples (Montoya et al. 2013) or directly in situ on sculptures (Angelini et al. 2011), but there is no agreement about the possibility of inferring a difference between corroding and noncorroding regions.

Aramendia used Raman spectroscopy to evaluate the composition of rust layers to calculate a protective ability index (PAI), which could be related to the stability and the aesthetic appearance of the rust layer (Aramendia et al. 2014b).

In order to ensure the longevity of structures, some aspects must be taken into account from the design stage until the moment when routine maintenance interventions are planned (*Construction Dimensions* 1980; ECCS 2001). During installation, it is important to remove mill scale, soil, and contaminants because their presence during the weathering process can disturb the formation of a uniform protective layer. When storage is necessary, the material should be placed in areas with good drainage. Contact with the ground, cover cloths, or plastics should be avoided.

The use of bare weathering steel is suited to most atmospheric environments (urban, rural, moderate industrial, and moderate marine). However, there are locations where the use of weathering steel in an uncoated condition is not recommended: in atmospheres containing concentrated corrosive industrial fumes, in marine locations subject to saltwater spray, or in applications where steel is continually submerged in water (salt or fresh) or buried in soil. In such conditions, the oxide layer will not be protective. The design of the artifact plays an important role; if water is allowed to flow over overlapping joints, capillary action can draw the water into the joint and cause “rust-pack” (exfoliation) to form.

It is also important to watch for moisture dripping from the steel, especially during its early years of exposure. It will contain particles of insoluble iron oxide, which can stain or streak adjacent materials—especially concrete, stucco, galvanized steel, stone, wood, and unglazed brick (Harris 2012). These conditions should be considered in the design, detailing, and the selection of proper adjoining materials.

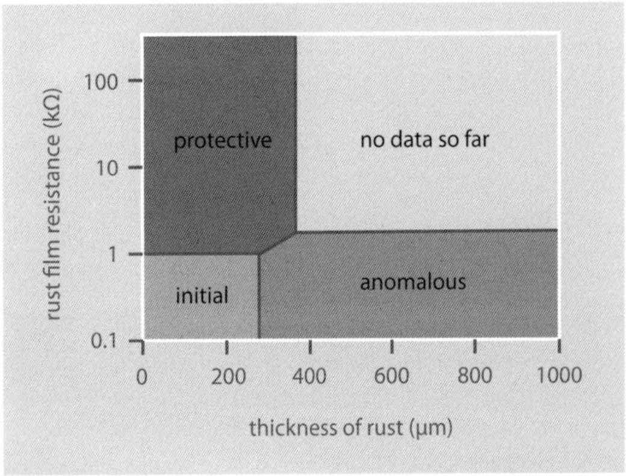

Figure 8.6. | Classification of rust conditions formed on weathering steel based on measurement of rust film resistance and thickness (adapted from Kihira, Ito, and Murata 1989, fig. 7).

CONSERVATION-RELATED LITERATURE

The corrosion behavior of weathering steel sculptures after several years of outdoor exposure was investigated (Angelini et al. 2011; Castro et al. 2011). Regarding intervention, a review of applicable techniques is described (Scott 1993). More specifically, procedures such as cleaning with dry ice have been used (Brush 2010; Maltby and Petrie 2016). Cathodic protection as preventive conservation (Ruiz de Arcaute 1998) and the utility of coatings were also evaluated (Decker, Brüggerhoff, and Eggert 2008; Schulz and Mai 2008).

REFERENCES AND BIBLIOGRAPHY

Angelini, E., S. Grassini, M. Parvis, and F. Zucchi. 2011. "An In-Situ Investigation of the Corrosion Behaviour of a Weathering Steel Work of Art." *Surface and Interface Analysis* 44:942–46.

Aramendia, J., L. Gomez-Nubla, I. Arrizabalaga, N. Prieto-Taboada, K. Castro, and J. M. Madariaga. 2013. "Multianalytical Approach to Study the Dissolution Process of Weathering Steel: The Role of Urban Pollution." *Corrosion Science* 76:154–62.

Aramendia, J., L. Gomez-Nubla, L. Bellot-Gurlet, K. Castro, C. Paris, P. Colomban, and J. M. Madariaga. 2014b. "Protective Ability Index Measurement through Raman Quantification Imaging to Diagnose the Conservation State of Weathering Steel Structures." *Journal of Raman Spectroscopy* 45 (11–12): 1076–84.

Aramendia, J., L. Gómez-Nubla, K. Castro, J. M. Madariaga, and L. Bellot-Gurlet. 2014a. "The Effect of Calcium Oxalates in the Weathering Steel Surface." In *Science, Technology and Cultural Heritage*, edited by M. A. Rogerio-Candelera, 211–16. London: Taylor and Francis Group.

ArcelorMittal. 2017a. "Indaten®: A Magical Steel, in Harmonious Dialogue with Nature." ArcelorMittal Europe Communications. *http://industry.arcelormittal.com/news/newsarchive/news2012/dec2012/newsweatheringsteel.*

——. 2017b. "Weathering Steel: For Unpainted Bridges and General Construction." ArcelorMittal USA. *http://www.usa.arcelormittal.com/~/media/Files/A/Arcelormittal-USA-V2/what-we-do/steel-products/plate-products/Weathering-steel.pdf.*

Asami, K., and M. Kikuchi. 2003. "In-Depth Distribution of Rusts on a Plain Carbon Steel and Weathering Steels Exposed to Coastal-Industrial Atmosphere for 17 Years." *Corrosion Science* 45:2671–88.

ASTM International. 2013. *ASTM A242-13. Standard Specification for High-Strength Low-Alloy Structural Steel*. West Conshohocken, PA: ASTM International.

——. 2015a. *ASTM A588-15. Standard Specification for High-Strength Low-Alloy Structural Steel, up to 50 ksi [345 MPa] Minimum Yield Point, with Atmospheric Corrosion Resistance*. West Conshohocken, PA: ASTM International.

——. 2015b. *ASTM G101-15. Standard Guide for Estimating the Atmospheric Corrosion Resistance of Low-Alloy Steels*. West Conshohocken, PA: ASTM International.

Brush, M. B. 2010. "Using Dry Ice for Spray-Paint Removal on Weathering Steel." *APT Bulletin* 41 (1): 1–5.

Castro, K., J. Aramendia, L. Gomez-Nubla, D. Vega, A. Sanz López de Heredia, A. García Ibáñez de Opakua, and J. M. Madariaga. 2011. "The Conservation of Weathering Steel Sculptures: A Comparative Study of Chillida's Sculptures Exposed in Bilbao." In *Art 2011: 10th International Conference on Non-Destructive Investigations and Microanalysis for the Diagnostics and Conservation of Cultural and Environmental Heritage, Florence, Palazzo degli Affari, FirenzeFiera 2011, April 13th–15th*. Florence: Istituto Superiore per la Conservazione ed il Restauro. *http://www.ndt.net/article/art2011/papers/GARCIA%20-%20E%2016.pdf.*

Chiavari, C., E. Bernardi, C. Martini, F. Passarini, A. Motori, and M. C. Bignozzi. 2012. "Atmospheric Corrosion of Cor-Ten® Steel with Different Surface Finish: Accelerated Ageing and Metal Release." *Materials Chemistry and Physics* 136:477–86.

Construction Dimensions. 1980. "Whiter Weathering Steel?" *Construction Dimensions* 13 (8).

Decker, P., S. Brüggerhoff, and G. Eggert. 2008. "To Coat or Not to Coat? The Maintenance of Cor-Ten® Sculptures." *Materials and Corrosion* 59 (3): 239–47.

Deflorian, F., and S. Rossi. 2002. "Premature Corrosion Failure of Structural Highway Components Made from Weathering Steel." *Engineering Failure Analysis* 9:541–51.

ECCS [European Convention for Constructional Steelwork]. 2001. "The Use of Weathering Steel in Bridges." Brussels: ECCS. *http://www.stalforbund.no/uploads/source/files/Ressurser/ECCS_No081_Corten_in_bridges.pdf.*

Harris, E. A. 2012. "Chic, Rusty Steel Facades Leave a Fiery Stain Below." *New York Times*, August 28, 2012.

Kihira, H. 2007. "Systematic Approaches toward Minimum Maintenance Risk Management Methods for Weathering Steel Infrastructures." *Corrosion Science* 49:112–19.

Kihira, H., A. Imai, M. Hiramatsu, T. Aiga, T. Matsumoto, T. Sato, and M. Nagai. 2009. "Additions to Contingency Plan for Enhancing Secureness of Weathering Steel Structures." *ECS Transactions* 16 (43): 115–24.

Kihira, H., S. Ito, and T. Murata. 1989. "Quantitative Classification of Patina Conditions for Weathering Steel Using a Recently Developed Instrument." *Corrosion* 45 (4): 347–52.

Kimura, M., H. Kihira, N. Ohta, M. Hashimoto, and T. Senuma. 2005. "Control of Fe(O,OH)6 Nano-Network Structures of Rust for High Atmospheric-Corrosion Resistance." *Corrosion Science* 47:2499–509.

Leygraf, C., and T. Graedel. 2000. "Weathering Steel." In *Atmospheric Corrosion*, 157–61. New York: Wiley InterScience.

Maltby, S. L., and L. Petrie. 2016. "The Conservation of Richard Serra's 'Tilted Spheres': When Public Art Becomes Too Public." In *Metal 2016: Proceedings of the Interim Meeting of the ICOM-CC Metals Working Group, September 26–30, New Delhi, India*, edited by R. Menon, C. Chemello, and A. Pandya, 292–97. Paris: ICOM-CC.

Misawa, T., K. Asami, K. Hashimoto, and S. Shimodaira. 1974. "The Mechanism of Atmospheric Rusting and the Protective Amorphous Rust on Low Alloy Steel." *Corrosion Science* 14:279–89.

Misawa, T., T. Kyuno, W. Suetaka, and S. Shimodaira. 1971. "The Mechanism of Atmospheric Rusting and the Effect of Cu and P on the Rust Formation of Low Alloy Steels." *Corrosion Science* 11:35–48.

Montoya, P., I. Díaz, N. Granizo, D. de la Fuente, and M. Morcillo. 2013. "A Study on Accelerated Corrosion Testing of Weathering Steel." *Materials Chemistry and Physics* 142:220–28.

Morcillo, M., B. Chico, I. Díaz, H. Cano, and D. de la Fuente. 2013. "Atmospheric Corrosion Data of Weathering Steels: A Review." *Corrosion Science* 77:6–7.

Morcillo, M., I. Díaz, B. Chico, H. Cano, and D. de la Fuente. 2014. "From Empirical Development to Scientific Design: A Review." *Corrosion Science* 83:6–31.

Nippon Steel and Sumimoto Metal Corp. 2017. *COR-TEN®*. *http://www.nssmc.com/product/catalog_download/pdf/006en.pdf.*

Pourbaix, M., and L. Miranda. 1976. "On the Nature of the Rust Layers Formed on Steels in Atmospheric Corrosion as a Function of Alloy Composition, Environmental Composition, Temperature and Electrode Potential." In *Passivity and Its Breakdown on Iron and Iron Base Alloys: U.S.A.-Japan Seminar*, edited by R. W. Staehle and H. Okada, 47–48. Houston: National Association of Corrosion Engineers.

Qian, Y. H., D. Niu, J. J. Xu, and M. S. Li. 2013. "The Influence of Chromium Content on the Electrochemical Behavior of Weathering Steels." *Corrosion Science* 71:72–77.

Ramirez-Barat, B., T. Palomar, B. García, D. de la Fuente, and E. Cano. 2016. "Composition and Protective Properties of Weathering Steel Artificial Patinas for the Conservation of Contemporary Outdoor Sculpture." In *Metal 2016: Proceedings of the Interim Meeting of the ICOM-CC Metals Working Group, September 26–30, New Delhi, India*, edited by R. Menon, C. Chemello, and A. Pandya, 314–19. Paris: ICOM-CC.

Ruiz de Arcaute, E. 1998. "Étude pour la protection d'une sculpture en corten: Possibilités et limites de la protection cathodique." In *Metal 98: Proceedings of the International Conference on Metals Conservation, Draguignan-Figanières, France, 27–29 May 1998*, edited by W. Mourey, 280–85. London: James & James.

Schulz, W.-D., and B. Mai. 2008. "Zur Korrosion und zum nachträglichen Korrosionsschutz von Industrie- und Kulturdenkmalen aus wetterfesten Baustählen." *Restauro* 7:456–71.

Scott, J. 1993. "Conservation of Weathering Steel Sculpture." In *Saving the Twentieth Century: The Conservation of Modern Materials. Proceedings of a Conference, Symposium '91: Saving the Twentieth Century, Ottawa, Canada, 15–20 September 1991*, edited by D. W. Grattan, 307–22. Ottowa: Canadian Conservation Institute.

———. 2014. "Weathering Steel." In *Twentieth-Century Building Materials: History and Conservation*, edited by T. C. Jester, 39–46. Los Angeles: Getty Conservation Institute.

Sheikh, N. M., and R. P. Bligh. 2013. *In-Field Inspection Methodology for Weathering Steel W-Beam Guardrail*. Study No. TPF-5(114). College Station: Texas A&M Transportation Institute, Roadside Safety Research Program.

Townsend, H. E., T. C. Simpson, and G. L. Johnson. 1994. "Structure of Rust on Weathering Steel in Rural and Industrial Environments." *Corrosion* 50 (7): 547–54.

WJE [Wiss, Janney, Elstner Associates]. 2013. *Assessment of Weathering Steel Bridge Performance in Iowa and Development of Inspection and Maintenance Techniques*. Report SPR 90-00-RB17-012. Ames: Iowa Department of Transportation Office of Bridges and Structures and Federal Highway Administration. *http://publications.iowa.gov/14956/1/Iowa_Weathering_Steel_Final_Report_2-21-2013.pdf*.

Wurth, L. 1991. "Role of Chromium in Weathering Steel Passivation." *Materials Performance* 3:62–63.

Yamashita, M., H. Konishi, J. Mizuki, and H. Uchida. 2004. "Nanostructure of Protective Rust Layer on Weathering Steel Examined Using Synchrotron Radiation X-rays." *Materials Transactions* 45 (6): 1920–24.

Yamashita, M., H. Miyuki, Y. Matsuda, H. Nagano, and T. Misawa. 1994. "The Long Term Growth of the Protective Rust Layer Formed on Weathering Steel by Atmospheric Corrosion During a Quarter of a Century." *Corrosion Science* 36 (2): 283–99.

CHAPTER 9 | ZINC ALLOYS

Cast and wrought alloys containing zinc as the major element constitute the third major use of zinc—after zinc coatings and zinc in brass (Porter 1991). Zinc-alloy castings containing aluminum are rich in design possibilities, because they can be produced to near net shape with useful mechanical properties. They are used in general hardware as well as in electronic and electrical fittings of all kinds. Zinc and its alloys show excellent bearing and wear properties because of their self-lubricating surface characteristics and excellent sound- and vibration-damping properties. In addition to their attractive properties, zinc alloys have a relatively low cost and are environmentally friendly.

Zinc-alloy castings were first produced for pressure die casting (metal injection into a mold under high pressure) at the beginning of the twentieth century and were initially made without regard to the purity of the zinc. The first alloys presented serious problems of swelling and disintegration, which have been attributed to intergranular corrosion stimulated by small amounts of impurities, such as lead, tin, and cadmium. Once the industry understood the necessity of metal purity, zinc-aluminum die casting alloys were increasingly developed and sold under different names, such as zamak (based on the first letters in German of zinc, aluminum, magnesium, and copper) and Mazak. In the 1970s, alloys with higher Al content (ZA) were developed, initially for gravity castings (Street 1985).

Wrought zinc alloys are produced in various shapes and compositions, depending upon the ultimate use of the product. The most widely employed alloys contain copper and titanium; they are used because they allow work hardening and improve the ability to withstand deformation under prolonged light loads (creep resistance).

METALLURGICAL BACKGROUND

Aluminum is the principal alloying element in the traditional zinc die casting alloys, providing good castability, which permits the production of thin-wall components. At room temperature, the solid solubility of aluminum in zinc is low—about 0.05 wt%—and the microstructure of the alloys consists of a mixture of zinc-rich or aluminum-rich dendrites with interdendritic eutectic structure (Davis 2001, 520–27).

Based on the aluminum content, casting alloys are divided into two principal groups (Caron and Staley 1997). Alloys with about 4% Al present zinc-rich dendrites, with a usual fairly fine equiaxed grain structure, which is primarily responsible for the strength, ductility, and toughness of zinc die castings. Alloys with >5% Al have aluminum-rich dendrites and increased strength, hardness, and stiffness, with lower ductility. Copper and magnesium are the other common alloying elements. Copper is added for strength, hardness, and creep resistance, and very low amounts of magnesium are added to counteract the harmful effects of impurities. In addition to the composition, the casting method will influence the cooling rate and therefore the resulting microstructure and properties (fig. 9.1).

Generally, wrought alloys have lower alloy content to allow working of the material. Because of the hexagonal close-packed (HCP) crystal structure of zinc, some properties of the rolled product are anisotropic (directional): the transverse orientation to the rolling direction shows higher tensile strength and lower thermal expansion than the rolling (longitudinal) orientation. Impurities with extremely limited solid solubility—especially lead, cadmium, and tin—must be tightly controlled to avoid microstructural weakness due to segregation (Porter 1991).

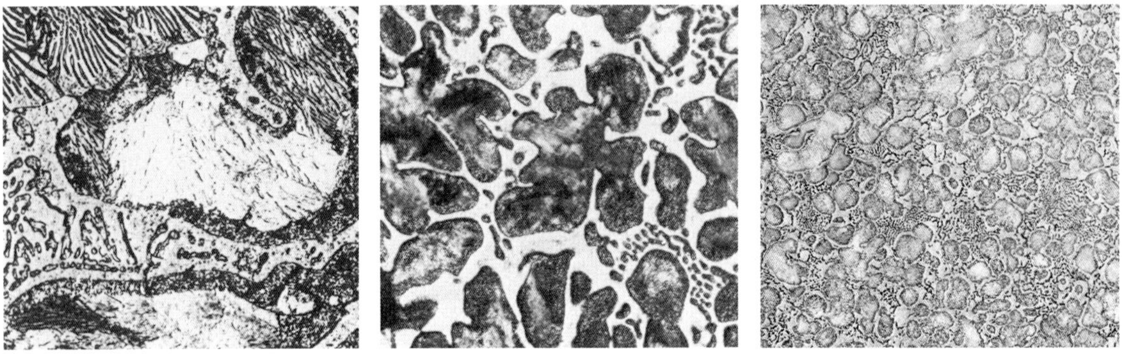

a b c

Figure 9.1 | Microstructure of ZA-12 alloy (UNS Z35631, Zn-11Al-0.9Cu-0.02Mg): Sand cast (a); permanent mold (b); and pressure die casting (c). All present zinc-rich dendrites, but they are finer for higher cooling rates (reproduced from Mongeon and Barnhurst 1985, figs. 7, 9, 11).

DESIGNATION, PROPERTIES, AND APPLICATIONS

As previously stated, zinc casting alloys can be divided into two basic families depending upon their aluminum content: Zamak alloys, containing about 4% Al, and ZA (zinc-aluminum) alloys, containing >5% Al (Barnhurst 1990). They are versatile engineering materials providing a unique combination of strength, toughness, rigidity, bearing performance, and economical castability. The principal casting alloys are presented in table 9.1.

With regard to wrought products, the unalloyed zinc used for roofing up to about the 1960s was soft and had a tendency to creep. As a consequence, it would eventually sag in the middle when the bays of fully supported roofing were too wide. The development of alloys with small additions of copper and titanium greatly reduced the tendency to creep. In addition to building components, another major use of zinc sheet is in dry cell battery production, which reached a peak in the mid-twentieth century.

Wrought zinc alloys are fabricated as flat-rolled, wiredrawn (used in thermal spraying, in which the wire is melted and sprayed onto a substrate with a special gun), and extruded and forged products. Examples of typical end uses of wrought zinc alloys are given in table 9.2.

Table 9.1. | Characteristics and applications of zinc-aluminum castings.

UNS (Common name)	Composition	Characteristics and applications
Z33520 (Zamak No. 3)	Zn-Al-Mg	Optimal combination of strength, castability, dimensional stability, ease of finishing, and cost
Z35531 (Zamak No. 5)	Zn-Al-Cu-Mg	Stronger and harder but less ductile than Zamak No. 3
Z33523 (Zamak No. 7)	Zn-Al-Mg-Ni	Improved casting fluidity, ductility, and surface finish; special hardware applications and for castings requiring extra formability during assembly operation
Z35541 (Zamak No. 2)	Zn-Al-Cu-Mg	The only zamak alloy used for gravity casting (known as Kirksite); mainly for metal forming dies or plastic injection tools
Z35635 (ZA-8)	Zn-Al-Cu-Mg	Combines high structural strength with excellent appearance (can be readily finished)
Z35630 (ZA-12)	Zn-Al-Cu-Mg	Combines casting qualities with optimal bearing and wear properties
Z35840 (ZA-27)	Zn-Al-Cu-Mg	Lightest, hardest, and strongest of all zinc alloys but has low ductility and impact strength; excellent sound and vibration damping properties

Source: Data from Barnhurst 1990.

Table 9.2. | Typical applications of wrought zinc alloys.

UNS	Composition	Applications
Z44330	Zn-Cu	Building construction materials, coinage, deep drawn hardware
Z41320	Zn-Cu-Ti	Roofing, gutters, solar collectors, address plates, deep drawn hardware, facade cladding
Z21540	Zn-Pb-Cd	Building construction materials, dry cell battery cans, electrical components, deep drawn hardware
—	Zn-Al (superplastic zinc)[a]	Shaped components such as computer panels and covers (e.g., three-storied pagoda in Tokyo[b])

Source: Data from Barnhurst 1990.

[a] The term *superplastic* as applied to metals refers to a condition in which the properties resemble those of the plastic state of thermoplastic polymers—that is, they can be uniformly and plastically elongated (exhibiting neither elastic behavior nor work hardening).

[b] Three-storied pagoda in the Chinzan-so Garden (Takenaka Corp. 2011).

Regarding surface finish, zinc alloys accept a wide assortment of coatings (Noranda 1992; IZA 2006). The coatings are generally specified to increase serviceability of the components and to provide long-term aesthetic properties. Coating systems available for zinc alloys are often classified into three categories: conversion, metallic, and organic.

Conversion finishes are usually based on the reaction between the alloy surface with proprietary chemical solutions, which form a complex conversion coating (see chap. 11). Chromate coatings provide low-cost corrosion protection that presents aesthetically pleasing colors. Phosphate conversion coatings are widely used primarily as a precoat for organic finishes on zinc castings. They improve adhesion, reduce paint blistering, and constitute an insoluble nonconductive barrier to the alloy, minimizing the spread of corrosion if the paint film is disrupted.

Zinc die castings can also be anodized prior to use or prior to painting. Through application of an adequate voltage, an anodic coating composed of oxides, phosphates, and chromates forms. Dense and porous, it is available in four colors: green, gray, charcoal, and brown. It can be produced over any zinc alloy and, because of its porosity, provides excellent paint adhesion and increased corrosion resistance after painting.

Metallic coatings on zinc castings are used for their appearance, corrosion resistance, and electrical properties (see chap. 10). For the most part, they are electrodeposited, and as a general rule, plate quality decreases with increasing aluminum content of the alloy. Electroless deposition is also applied, offering a more uniform coating thickness (especially for intricate forms), improved hardness, and wear resistance.

DETERIORATION

Zinc-based alloys have good resistance to corrosion in most natural environments owing to their ability to form a protective layer comprising carbonates, oxides, or hydrated sulfates. Once these layers have formed and completely cover the surface, corrosion proceeds at a greatly reduced rate. However, since zinc oxide is amphoteric, the formation and maintenance of the protective layer is governed largely by the pH of the environment. Thus, acidic pollutants and strong alkaline conditions adversely affect the corrosion behavior by interfering with the formation of the protective layer (Porter 1994).

Zinc-aluminum alloys usually oxidize in a uniform manner, producing dulling and graying of the finish with time. Prolonged or repeated contact with wet concrete, plaster, or other alkaline building materials, as well as with green woods and wood preservatives, causes corrosion.

As mentioned above, intergranular corrosion of zinc alloys was a problem in the past, because of the high impurity level—notably of lead, tin, and cadmium—in the unalloyed zinc (Goodway 1985). Figure 9.2 shows an example of corrosion-induced cracking in a zinc-aluminum die-cast model locomotive resulting from impurities in the metal.

The subsequent introduction of 99.99% zinc as the base for producing alloys eliminated intergranular corrosion under normal service conditions. In practice, the use of zinc-aluminum casting alloys should be restricted to temperatures below 70°C, and impurities should be kept under specific limits (Zhang 1996, 227–34; Wanhill and Hattenberg 2005).

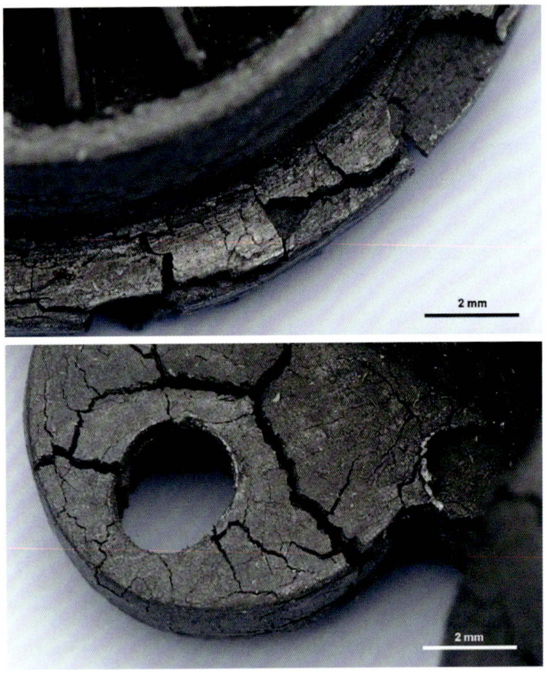

a

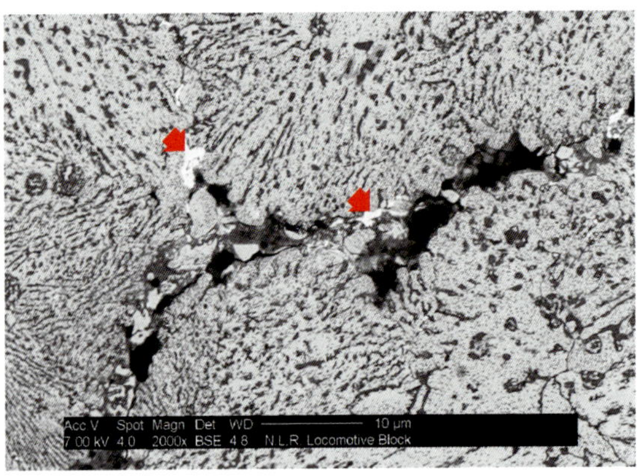

b

Figure 9.2 | Severe corrosion-induced cracking observed in a zinc-aluminum die-cast model train (a); scanning electron microscope (SEM) image showing corrosion products filling a crack and lead particles (arrows) along crack patterns (b) (reproduced from Wanhill and Hattenberg 2005, figs. 1, 9).

CHARACTERIZATION

Even though they are often finished with a number of different coatings (metallic, chromate, painted, etc.), Zn alloy die castings can be more or less recognized by the type of artifact, considering both form and use. Because of their unique combination of properties—including excellent castability, low melting point, and low cost—these alloys are preferred for their ability to allow inexpensive production of series of artifacts with intricate forms. The inconvenience they present is their fragility: they break easily when subjected to impact, leaving the fracture section with a granular aspect (fig. 9.3).

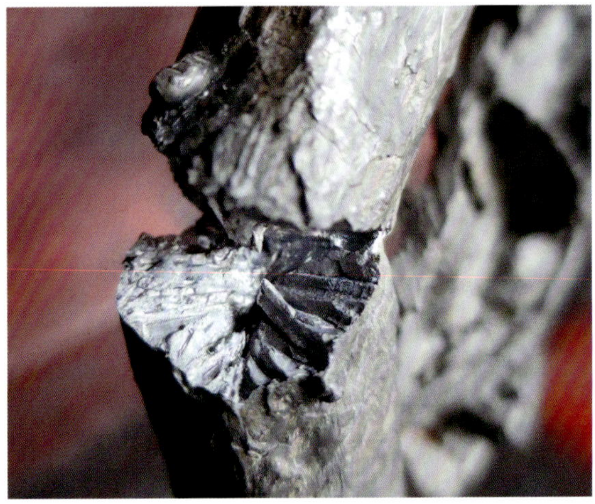

Figure 9.3 | Granular aspect of the fracture of a cast zinc alloy. Photo: Courtesy of Antoine Amarger.

CONSERVATION-RELATED LITERATURE

Publications about zinc alloys are for the most part related to the conservation of sculptures, most of them situated outdoors (Weil 1987; Grissom 1993), especially in Germany (Riederer 1993, 1994), the United States (Grissom 2009, 2012; Grissom and Harvey 2003), and France (Texier 2010). Some papers deal with specific problems, such as intergranular corrosion (Mottner, Assfalg, and Freitag 1998) and the use of various organic coatings to protect the artifacts (Mottner, Freitag, and Brückner 1995; Mottner, Brückner, and Freitag 1997).

Another topic generally published is the conservation of zinc elements of monuments (Hermans and van Hunen 2003; Gayle and Look 1992; Takenaka Corp. 2011), with a special focus on the particularities of zinc roofing (Pieper 1990; English Heritage 2013; Hermans and van Hunen 2003).

Other objects of investigation have been the conservation of zinc plates used for engraving (Lage de la Rosa 2003; Lage and Mota 2008; Zappalà 1990a, 1990b) and zinc plates used as supports for paintings (Thistlewood and Northover 2009).

Two publications address broader themes: the general care of zinc objects (Barclay 2007) and the effect of museum pollutants on zinc-alloy conservation (Mosca, Camões, and Casanovas 2007).

REFERENCES AND BIBLIOGRAPHY

Barclay, R. L. 2007. *Care of Objects Made of Zinc—Canadian Conservation Institute (CCI) Notes 9/9.* Ottawa: Canadian Conservation Institute. *https://www.canada.ca/en/conservation-institute/services/conservation-preservation-publications/canadian-conservation-institute-notes/care-zinc.html.*

Barnhurst, R. J. 1990. "Zinc and Zinc Alloys." In *Properties and Selection: Nonferrous Alloys and Special-Purpose Materials*, 527–42. Vol. 2 of *ASM Handbook*. Materials Parks, OH: ASM International.

Caron, R. N., and J. T. Staley. 1997. "Zinc and Zinc Alloys." In *Materials Selection and Design*, 383–415. Vol. 20 of *Effects of Composition, Processing, and Structure on Properties of Nonferrous Alloys*. Materials Park, OH: ASM International.

Davis, J. R. 2001. "Zinc and Zinc Alloys." In *Alloying: Understanding the Basics*, edited by J. R. Davis, 520–27. Materials Park, OH: ASM International.

English Heritage [Historic Buildings and Monuments Commission for England]. 2013. "Zinc." In *Practical Building Conservation: Roofing*, 493–512. Farnham, UK: Ashgate.

Gayle, M., and D. L. Look. 1992. "Zinc." In *Metals in America's Historic Buildings*, edited by M. Gayle and J. G. Waite, 15–21. Washington, DC: National Park Service.

Goodway, M. 1985. "Disintegration of Zinc Die-Castings." *Historical Metallurgy* 19 (1): 37–38.

Grissom, C. A. 1993. "The Conservation of Outdoor Zinc Sculpture." In *Ancient and Historic Metals: Conservation and Scientific Research: Proceedings of a Symposium Organized by the J. Paul Getty Museum and the Getty Conservation Institute, November 1991*, edited by D. A. Scott, J. Podany, and B. Considine, 279–304. Los Angeles: Getty Conservation Institute.

———. 2009. *Zinc Sculpture in America, 1850–1950*. Newark: University of Delaware Press.

———. 2012. "The Zinc Statuettes of Cornelius and Baker." *Winterthur Portfolio* 46 (1): 25–61.

Grissom, C. A., and R. S. Harvey. 2003. "The Conservation of American War Memorials Made of Zinc." *Journal of the American Institute for Conservation* 42 (1): 21–38.

Hermans, T., and M. van Hunen. 2003. "Bladzink op monumenten." *RDMZ Info* 35:1–8.

IZA [International Zinc Association]. 2006. "A Performance Evaluation of Modern Surface Finishes for Zinc Die-Castings." Brussels: International Zinc Association.

Lage, M., and J. M. Mota. 2008. "Characterization of Zinc Engraving Plates in Contemporary Art as an Aid in Their Conservation-Restoration." In *15th Triennial Conference, New Delhi: 22–26 September 2008*, vol. 1, 465. New Delhi: Allied Publishers.

Lage de la Rosa, M. 2003. "Caracterización del zinc calcográfico contemporáneo para su conservación y restauración." PhD diss., Universidad Complutense de Madrid.

Mongeon, L., and R. J. Barnhurst. 1985. "Zinc and Zinc Alloys." In *Metallography and Microstructures*, 9th ed., edited by ASM Handbook Committee, 488–96. Vol. 9 of *Metals Handbook*. Metals Park, OH: American Society for Metals.

Mosca, A. I., M. F. Camões, and L. E. Casanovas. 2007. "Materiais poluentes e controlo da poluição no Museu da Fundação Cupertino de Miranda, Porto." *Conservar Património* 6:31–37.

Mottner, P., E. Assfalg, and J. Freitag. 1998. "Investigations into Intercrystalline Corrosion and Conservation of Zinc." In *Metal 98: Proceedings of the International Conference on Metals Conservation, Draguignan-Figanières, France, 27–29 May 1998*, edited by W. Mourey and L. Robbiola, 329–40. London: Earthscan.

Mottner, P., H. Brückner, and J. Freitag. 1997. "Conservation of Cast Zinc Sculptures in Outdoor Exposure." In *Metal 95: Actes de la conférence internationale sur la conservation des métaux = Metal 95: Proceedings of the International Conference on Metals Conservation: Semur en Auxois, 25–28 September 1995*, edited by I. D. MacLeod, S. L. Pennec, and L. Robbiola, 251–55. London: Earthscan.

Mottner, P., J. Freitag, and H. Brückner. 1995. "Condition and Conservation of 19th Century Outdoor Zinc Sculpture." In *From Marble to Chocolate: The Conservation of Modern Sculpture—Tate Gallery Conference, 18–20 September 1995*, edited by J. Heuman, 39–45. London: Archetype Publications.

Noranda. 1992. "Designing Zinc Castings for Corrosion Resistance." Toronto: Noranda Sales Corp.

Pieper, R. 1990. *Restoring Metal Roof Cornices*. Preservation Tech Notes, Metals, no. 2. Washington, DC: National Park Service.

Porter, F. 1991. *Zinc Handbook: Properties, Processing, and Use in Design*. New York: Marcel Dekker.

———. 1994. *Corrosion Resistance of Zinc and Zinc Alloys*. New York: Springer Science+Business Media.

Rabinowitz, M., and C. Grissom. 2014. "Conservation of American Zinc Sculpture: History of the Use of the Material and Case Studies of Treatment." In *Open Air Metal: Outdoor Metallic Sculpture from the XIXth to the Early XXth Century*, edited by SFIIC [Section française de l'Institut International de Conservation], A. Azema, and A. Texier, 97–104. Paris: ICOMOS France.

Riederer, J. 1993. "Corrosion Damage to Zinc Sculptures in Germany." *European Cultural Heritage Newsletter on Research* 7 (1–4): 75–81.

———. 1994. "Schäden an Zinkskulpturen im Freien. Bisherige Erhaltungmaßnahmen, bisher eingesetzte Materialien: Wie sind sie zu werten?" *Restauro: Forum für Restauratoren, Konservatoren und Denkmalpfleger* 100 (4): 236–43.

Street, A. 1985. "Zinc: Zinc Alloys, 1900–1950." *Historical Metallurgy* 19 (1): 31–37.

Takenaka Corp. 2011. "Completion of the 'Heisei Refurbishing Work' on the "Three-Storied Pagoda in the Chinzan-so Garden,' One of Three Old Pagodas Existing in Tokyo." *http://www.takenaka.co.jp/takenaka_e/about/news/2011/m1001_06.html.*

Texier, A. 2010. "La sculpture monumentale en métal: Le zinc dans la sculpture monumentale." *Monumental*, semestriel 2:96–97.

Thistlewood, J., and P. Northover. 2009. "Corrosion Analysis and Treatment of Two Paintings on Zinc Supports by Frederick Preedy." *Journal of the Institute of Conservation* 32 (2): 137–48.

Wanhill, R. J. H., and T. Hattenberg. 2005. "Corrosion-Induced Cracking of Model Train Zinc-Aluminium Die Castings." Report NLR-TP-2005-205. Amsterdam: National Aerospace Laboratory.

Weil, P. D. 1987. "Problems in the Conservation of Zinc Sculpture in Outdoor Exposure." In *Conservation of Metal Statuary and Architectural Decoration in Open-Air Exposure: Symposium, Paris, 6–8 October 1986*, 157–68. Rome: ICCROM.

Zappalà, A. 1990a. "Indagini e proposte per il restauro conservativo delle matrici calcografiche." In *Morandi. L'opera gráfica. Rispondenze e variazioni*, edited by M. Cordaro, 61–66. Milan: Electa.

———. 1990b. "Pulizia dai prodotti di corrosione matrici calcografiche di zinco." *Kermes* 8:18–26.

Zhang, X. G. 1996. *Corrosion and Electrochemistry of Zinc*. New York: Springer Science+Business Media.

PART III | SURFACE COATINGS

CHAPTER 10 | METALLIC COATINGS

The surface of an artifact is the interface with its surroundings; it is what one sees and touches, and so it has considerable influence on the viewer's experience. The surface determines the response to chemical and physical circumstances, its resistance to tarnishing and corrosion, and its tolerance to impact and abrasion. The use of metallic coatings enables the bulk properties of one material to be combined with the surface properties of another. In the case of metallic substrates, coatings are applied to improve certain characteristics, such as corrosion resistance (e.g., by offering a barrier or sacrificial protection), wear resistance, electronic or electrical properties, and aesthetic appearance.

The surface finishing industry was strongly impacted by the development of electric generators in the late 1800s. Since that time, numerous plating processes have been developed, and after World War II, processes such as electroless plating, electropolishing, vacuum metallizing, mechanical plating, electrostatic painting, electrolytic painting, and powder coating were increasingly commercialized.

From the whole surface coatings family—which includes organic and inorganic finishes—only two types of inorganic coatings will be addressed in the present volume: metallic coatings and conversion coatings (chap. 11). Metallic coatings result from adding an extra metallic layer (i.e., a metal overlay) on the surface. Processes include electroplating, electroless plating, and hot-dip galvanizing.

COATING PROCESSES

Some of the numerous processes employed to deposit metallic coatings are introduced below.

Electroplating

Electroplating is the process of depositing a metallic coating on a surface by means of electrolysis. Since electroplating requires the passage of an electrochemical current, its beginning dates to the discovery of electricity (Watt and Phillip 2005; Raub 1993; Pavlova 1963; Flandro and Loveman 2016).

Shortly after Alessandro Volta invented the first battery, in 1803 one of his colleagues, Luigi Brugnatelli, used it as a source of current to electrodeposit gold on two silver medals. A quarter century passed before Michael Faraday defined the laws of electrolysis, and in 1836 John Frederic Daniell developed the cell named after him—a much more reliable source of current than previously existing ones. Such cells were installed by the Elkington cousins, who established the electroplating of silver and gold on an industrial scale in 1839. Nickel plating was first performed by G. Bird in 1837, but it was not until 1868 that Isaac Adams Jr. was able to put the process into commercial operation. Zinc and copper electrodeposition were developed over the same period. A period of over fifty years ensued before chromium plating emerged in the 1920s. Today the field of commercial electrodeposited metallic coatings embraces a lot of different metals and alloys (Clauss 1974).

A plating cell consists of two electrodes dipped into an electrolyte—that is, a conductive solution consisting of ionic species dissolved in a solvent, usually water. The electrodes are connected externally to the poles of a power supply. In the simplest case of copper deposition, the electrolyte consists of a copper sulfate solution, which dissociates easily in water, giving positive ions (Cu^{2+}) and negative ions (SO_4^{2-}). One of the two electrodes is the cathode, the part to be covered with copper; it is connected to the negative pole of the power supply. The other electrode is the anode, usually

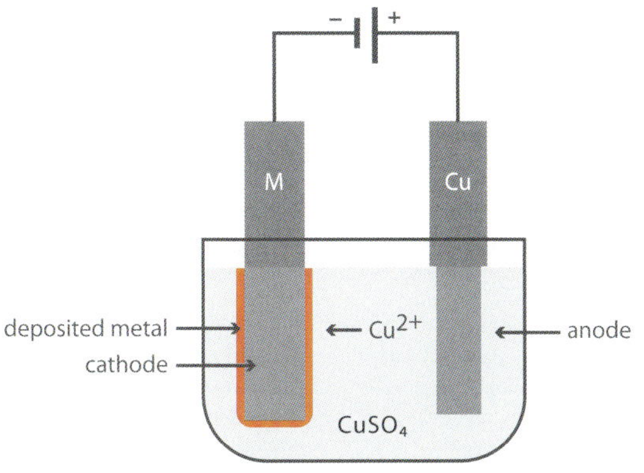

Figure 10.1 | Electrodeposition of copper from a copper sulfate solution: Reduction of positive ions takes place at the cathode, producing a metallic copper layer, while the anode dissolves, providing ions to the electrolyte.

of the same metal to be deposited, which is connected to the positive pole (fig. 10.1).

Electroplating is performed by passing an electrical current between the two electrodes. The electrical current delivered by the power supply is of an electronic nature, whereas in solution it results from the migration of ionic species between the two electrodes: positive species move toward the cathode and negative toward the anode, closing the circuit. The net reaction is

$$Cu^{2+} + 2e^{-} \rightarrow Cu(0)$$

where Cu(0) is a metallic copper atom.

In order to obtain good-quality coatings, several matters—such as surface preparation, electrolyte constitution, and operational conditions—must be carefully addressed (Edwards 1983).

Surface preparation is critical for obtaining good adhesion of the coating on the substrate: the base metal must be clean and free of grease and oxide films. Preparatory treatments generally include several stages with intermediate rinses and may differ substantially according to the nature of the substrate.

Electrolyte constitution must be considered. The electrodeposition of metal coatings is usually based on aqueous electrolytes, known as *baths*. Their primary constituent is the metallic salt of the metal to be deposited, whose solubility is often restricted to given pH values. Among the simple salt baths or acid baths, sulfate metallic salts are the most commonly used, often mixed with chlorides, tartrates, or carbonates. Complexing agents, such as cyanides, allow the use of alkaline baths. Moreover, a number of additives are necessary to optimize the deposit properties. They are: (1) wetting agents, which help with detachment of bubbles and eliminate pitting; (2) conducting salts and buffers, to promote charge transfer and avoid local pH changes; (3) addition agents: small amounts of certain substances, often organic, that may affect surface finish or the microstructure (grain size) of deposits, so as to obtain a brighter or leveled surface or to reduce internal stress.

Operational conditions are another important factor: One may prefer applying either a fixed current or a fixed voltage. The bath agitation and the geometry of the part are also critical parameters that control the growth rate and uniformity of the coating. Together with the factors mentioned above, they determine two important characteristics of the plating system. The first one is its *efficiency*. For instance, silver is deposited from a cyanide electrolyte with 100% cathode efficiency—which means that all of the applied current is used to reduce silver at the cathode surface. By comparison, chromium electroplating from an acid-sulfate bath has only 15% efficiency because 85% of the total current is consumed in other cathodic processes, such as hydrogen gas evolution. The second important characteristic of a plating system is its *throwing power*, which is its ability to produce deposits of uniform thickness on surfaces with irregularities (hollows, grooves, edges) and on large-scale profiles. The geometry of the plating cell arrangement (fig. 10.2) is crucial, as is the conductivity of the bath.

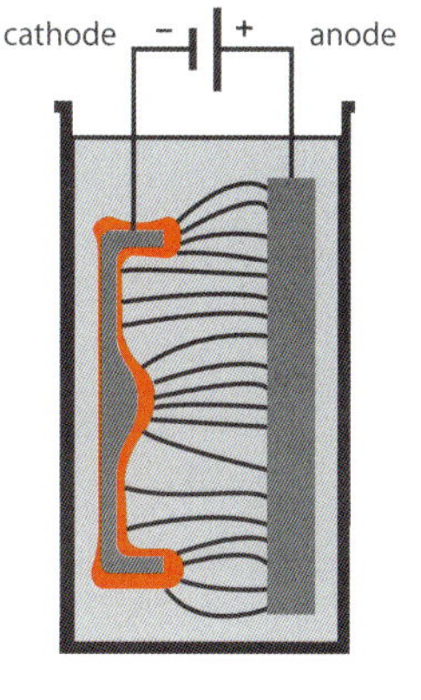

Figure 10.2 | Typical distribution of a deposit: Thickening at the edges and thin deposits in the center of concave surfaces (adapted from Nickel Institute 2014, fig. 2).

Moreover, some electrolytes are able to produce more or less leveled deposits. *Leveling* is defined as the progressive reduction of the surface roughness during deposition: more metal is deposited on recessed areas than on the protruding areas (fig. 10.3).

For the most commonly electroplated metals, these important characteristics are listed in table 10.1. It is noteworthy that they can differ considerably for the same electrodeposited metal, depending on the nature of the electrolyte. This information can be helpful in identifying the probable type of bath used to coat a given part.

Besides the traditional setup depicted in figure 10.1, electroplating can be performed in alternative ways: continuous plating, brush (selective) plating, alloy plating, plating on plastics, and electroforming. These methods are briefly described below.

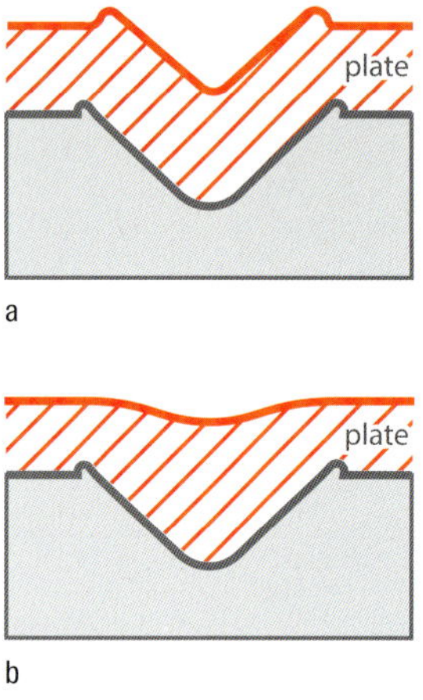

Figure 10.3 | Nickel deposit profiles: Absence of leveling (a); leveling provided by a leveling agent (b) (adapted from Nickel Institute 2014, fig. 3).

Table 10.1. | Process characteristics and applications of commonly electroplated metals.

Metal or alloy	Bath type	Plating rate (efficiency)	Throwing power (covering complex shapes)	Major applications
Cd	Alkaline/cyanide	Medium	Good	Functional
Cr	Acid/hexavalent	Low	Poor	Decorative, functional
	Slightly acid/trivalent	Low	Good	Decorative
Cu	Acid/sulfate	Medium	Poor/fair	Functional, leveling
	Alkaline cyanide	Medium	Good	General undercoat
	Slightly alkaline/pyrophosphate	Medium	Fair	Functional, undercoat
Cu-Zn	Alkaline cyanide	Low	Good	Decorative, adhesion to rubber
Au	Slightly alkaline	Low	Good	Decorative
	Slightly acid	Low	Good	Electrical, decorative
Pb	Acid/fluorborate	High	Good	Protective (sulfuric acid)
Pb-Sn	Acid/fluorborate	Medium	Good	Lead bearings, high-tin solder
Ni	Slightly acid Watts	Medium	Fair	Decorative, functional
	Slightly acid sulfamate	High	Fair	Functional
Ag	Alkaline cyanide	Medium	Good	Decorative, functional
Sn	Alkaline stannate	Medium	Good	Protective, electrical
	Acid/sulfonic	High/Medium	Fair	Continuous; functional and decorative
	Acid/fluorborate	High	Good	Wire bearings
Zn	Alkaline cyanide	Medium	Good	Protective, decorative
	Alkaline zincate	Low	Good	General
	Acid sulfate	Medium	Fair	Wire cast and wrought iron

Source: Data from Edwards 1983, 49–51.

Continuous Plating

Continuous electrodeposition involves electroplating a pure metal or alloy coating onto an endless steel strip (arbitrarily defined as having a rectangular cross section with a maximum thickness of 5 mm). In practice, this is achieved by passing a strip coil through the sequence of surface preparation, electroplating (fig. 10.4), and surface posttreatment. The facilities used for this process are commonly called continuous plating lines. The technology offers three main advantages: (1) high-speed continuous strip processing; (2) good control and uniformity of coating thickness, allowing the use of thinner coatings; and (3) the ability to plate different coating thicknesses on each surface of the strip, producing differential plates that can cost-effectively meet the different corrosion resistance requirements of the inside and outside of containers.

The metals most used for this kind of plating are zinc, tin (Smith 1978), chromium, and alloys of zinc with either nickel or iron. Zinc and zinc alloys, when continuously electroplated onto steel, is referred to as *electrogalvanized sheet* and used for automobile body panels, furniture, and appliances such as toasters, dishwashers, washing machines, and dryers.

Electrolytic tin-plated strip, commonly called tinplate, is used for applications requiring severe forming (for example, parts with excessively sharp radii or excessively hard work/low formability metal). The tin coating serves as a lubricant as well as provides corrosion protection. Tinplate has been used to produce toys, which were usually painted by a lithographic process (fig. 10.5).

The continuous plating of wire is another important industrial process. Particular applications are tin-coated copper wire for electrical use, zinc-plated steel wire for ropes and hawsers, and brass-coated steel wire for use in tires. Several other metallic coatings, such as copper, nickel, and terne (Pb-Sn), are also applied by continuous steel strip plating but on a relatively smaller scale and production volume.

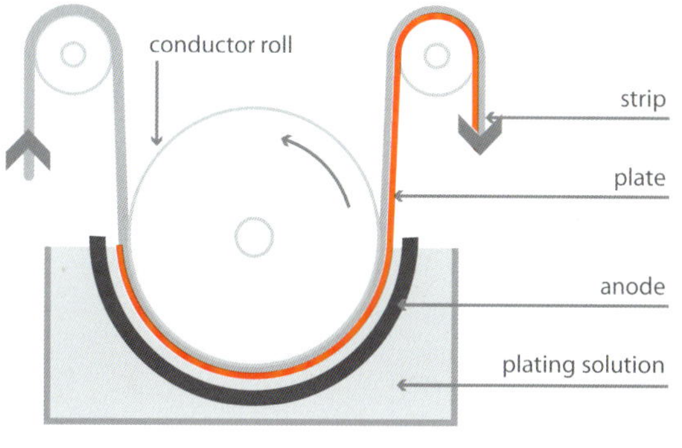

Figure 10.4 | A strip acting as a cathode passing through a plating solution.

Brush Plating

Brush plating—also known as selective plating—differs from traditional tank, or bath, plating in that the workpiece is not immersed in a plating solution (electrolyte). Instead, a handheld inert anode or stylus, covered with an absorbent wrapping and wetted with the electrolyte, is applied to the part (cathode) (Browning 1994). A direct current power pack drives the electrochemical reaction, depositing the desired metal on the substrate. In practice, movement between the anode and cathode is required for optimal results. A schematic of the brush plating process appears in figure 10.6.

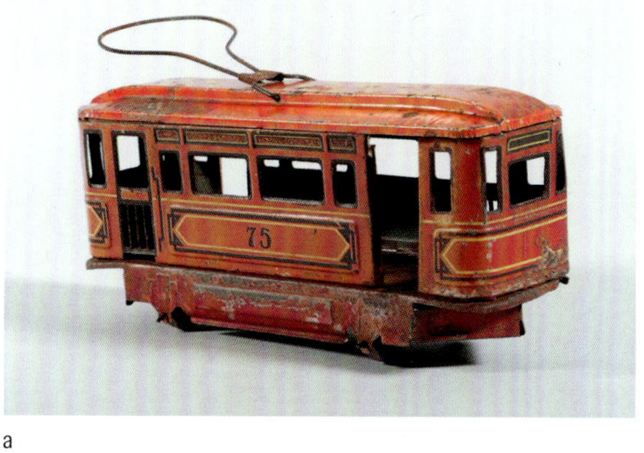

a

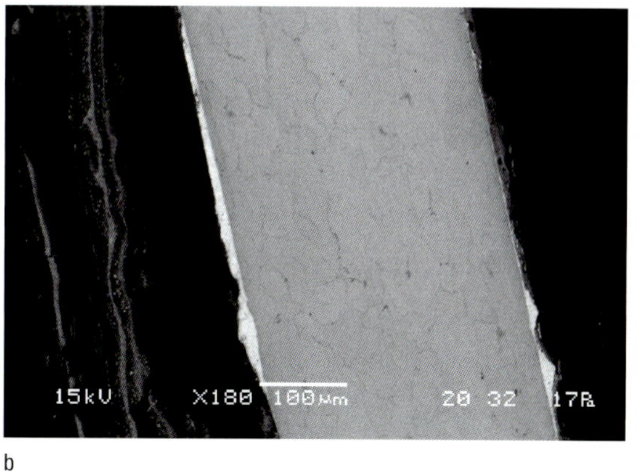

b

Figure 10.5 | Toy tinplate streetcar (Metalma collection, Museu Paulista, São Paulo) (a); cross section showing the steel strip covered by a relatively discontinuous tin layer (white) and the external paint layers (on the left) (b).

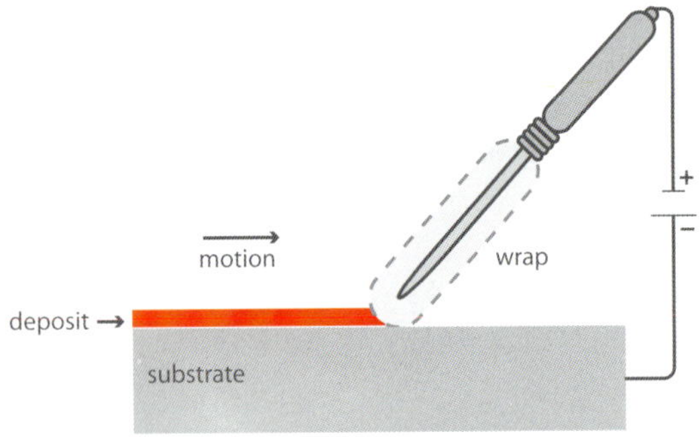

Figure 10.6 | Brush plating process: A wrapped inert anode, containing the electrolyte, is brushed onto the surface to be plated.

Materials used as the anode must be able to carry high current densities and be inert in plating solutions. Among them are graphite, which can be machined or shaped to fit the contour of the part being processed; stainless steel, which is more durable but dissolves in some plating solutions; and costly platinum alloys.

The anode covers (wrapping materials) serve to physically separate the anode and cathode and help ensure smooth deposits at high current densities. Because they hold the electrolyte, covers must be completely free of oil and foreign materials. Various fiber-like materials make suitable covers. Cotton works very well, and long-fiber synthetic fibers, such as polyester and nylon, although they do not wet or hold electrolytes as well as cotton, can be used. Most of these materials work well in the form of tube gauze ("bandage covers") as covers over cotton.

The key advantage of brush plating is its portability. It can be used to repair and restore dimensions or to apply a wear-resistant or corrosion-resistant surface to any selected area of a larger part (Sifco 2017).

Alloy Plating

The codeposition of two or more metals requires that their reduction potentials be fairly close to each other in a given electrolyte. This can be achieved either by adjusting the relative concentrations of the metals or by using complexing agents to control the free metal ion available to the reaction (Faust 1974). Numerous metals can be codeposited, producing alloys (fig. 10.7).

Brass plating is one of the most common alloy plating processes. All alloys from pure zinc to pure copper can be plated, with results matching the corresponding wrought alloys in color and properties. Because of its decorative effect, electroplated brass over aluminum and zinc base die castings, steel tubing, and stampings is widely used as a substitute for solid brass to reduce weight and cost. Brass plating is also employed for aesthetic reasons, such as to mimic gold finishes for cosmetic items (golden-colored brasses, 85%–90% Cu) or, when followed by various processes, to produce antique or other dark finishes. Brass plate has been used in musical instruments as well as in tire cod wires, providing the best adhesion of rubber to steel (Geissman and Bennet 1974).

Tin and copper can be codeposited in all proportions. Alloys containing 10%–15% Sn are attractive and used for decorative wares. They have a gold color that is, however, browner than true gold. Bronze plating is used when improved lubricity and wear resistance against steel are desired. The 45% Sn alloy is known as speculum metal; it has a silvery appearance, and in the past, it was used for decorative indoor items (Strow 1994).

The development of several types of iron-alloy plating is one of the most important achievements of the metal finishing industry. Iron-nickel-alloy plating is widely used for decorative coatings, for the surface coating of continuous

	Au	Cr	Cu	Fe	Ni	Pb	Pt	Sn	Zn
Sn									●
Pt									
Pb							●	● (Zn)	●
Ni						●	●	●	●
Fe					● (W)	●		●	●
Cu				● (Ni)	● (Zn)(Sn)	● (Sn)	●	● (Zn)	●
Cr			● (Zn)		●			● (Zn)	●
Au		●	●		● (Sn)		●	●	
Ag	● (Cu)	●	● (Pb)	●	●	●	● (Au)	● (Au)	● (Pt)(Au)

Figure 10.7 | Alloys that have been electrodeposited from aqueous solution. Binary alloys are indicated by orange dots at the intersection of a column and a line headed by a parent metal; ternary alloys are indicated by parentheses enclosing the symbol of the third element (adapted from Brenner 1963, fig. 1.2).

casting molds in the steel industry, and for magnetic devices. Iron-zinc-alloy plating is used in the steel industry for the corrosion protection of iron-based materials (Izaki 2000).

Plating on Plastics

Plating on plastics is a term used to describe the decorative or functional application of metal onto plastic substrates using the process of plating from an electrolyte. Although a wide range of plastic resins can be plated, the market has always been dominated by acrylonitrile butadiene styrene (ABS). Other resins used for plating include nylon, modified polyesters, polysulphones, polyimides, and polyetherimides.

Since plastics are electrically nonconductive, a special technique was developed to allow plating on plastics. In the mid-1960s, the development of etching techniques combined with advances in electroless plating gave rise to a system that provided a highly conductive coating exhibiting satisfactory adhesion to the plastic surface (fig. 10.8). After etching with an oxidizing agent, the roughened plastic surface is coated with traces of a precious metal, which provides the "seeds" for the further addition of a thin layer of nickel or copper by electroless deposition. As a top layer, nickel and chromium metals are the most commonly electroplated, in processes usually called chrome plating or plastic chrome plating (Kuzmic 1990).

This type of coating provides both technical and aesthetic benefits and can be applied to meet many different application criteria. For example, highly visible and corrosion-resistant exterior automotive components are often chrome-plated plastics, providing a lower-weight option in comparison with traditional metal components. Plastic chrome plating is also ideal for sanitary fittings that require a durable and wear-resistant coating to resist humid bathroom environments. Other significant uses include knobs and buttons for the electronics and white goods (domestic applicances) industry. In France in particular, the production of high-quality perfume bottle caps in very large numbers represents a very important sector of the market (Parkinson and Hart 1995). An example of art produced in this method is Robert Watts's *Chrome Cabbage*, in the collection of New York's Museum of Modern Art (fig. 10.9).

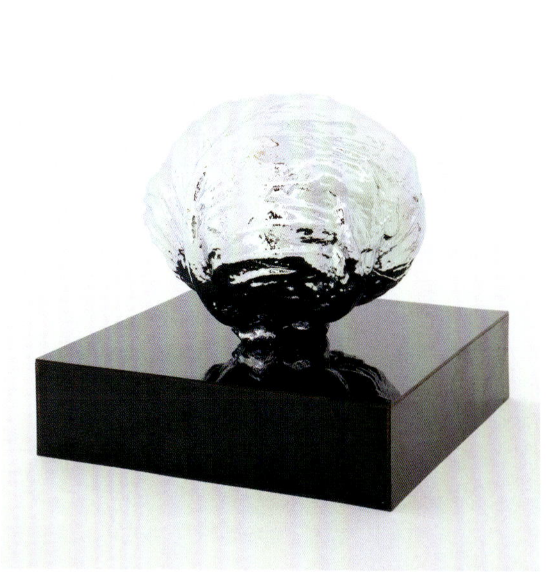

Figure 10.9 | Chrome plating over plastic cabbage. *Chrome Cabbage* by Robert Watts (1964), Museum of Modern Art, New York. Patricia Phelps de Cisneros, Emily Rauh Pulitzer and Richard S. Zeisler Funds.

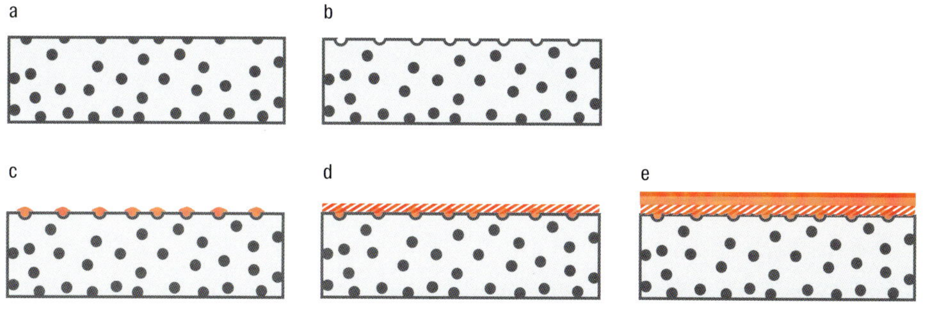

Figure 10.8 | Sequence for a typical process of plating on plastic: Untreated ABS plastic (a); etched surface (b); activated (c); with an electroless layer (e.g., Ni or Cu) (d); with one or more electroplated layers (e.g., Ni, Cr) (e).

Electroforming

Electroforming is the process by which articles or shapes can be reproduced exactly by electroplating on a model that can be later removed, leaving a precise duplicate of the original— called an *electrotype* or an *electroformed part* (fig. 10.10). Electrodeposition is performed until the plate itself is sufficiently thick and strong to be self-supporting. The most important attribute of electroforming is its ability to replicate patterns and textures accurately in a rigid and durable material.

Developed in 1838 by Moritz von Jacobi in Russia, the electroforming industry was initially limited by the sources of the electric current needed for the process. By the 1870s, with the advent of mechanical electric generators (dynamos), electroforming was widely adopted to produce printing plates. Electroforming has also been used to reproduce valuable objects, such as ancient coins, and for the production of copper sculptures, sometimes called *galvanoplastic bronzes*. Among the earliest documented electroform sculptures are Josef Hermann's twelve angels (1858) at the base of the cupola of Saint Isaac's Cathedral in St. Petersburg; the statue of Prince Albert (1863) in London; and Oudry's fountains in the Place de la Concorde, Paris (1860s) (Watt and Phillip 2005). From 1890 through at least 1930, the Abteilung für Galvanoplastik of the WMF company in Germany produced many statues and other items using electroforming (Haber and Heimler 1994). Moreover, since the end of the nineteenth century, electroforming has been used to produce record masters (fig. 10.11) and to replicate many well-known artworks for museums all over the world

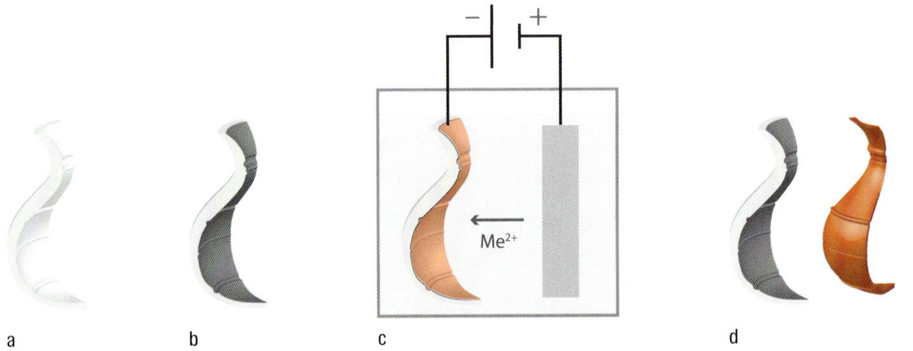

a b c d

Figure 10.10 | Typical copper electroforming sequence: A flexible mold material is produced after a model (a); a coating of graphite makes the interior of the mold electrically conductive (b); an electrical charge is applied to the mold in a copper sulfate bath, and a thick copper wall is eventually formed (c); the copper form is separated from the mold (d).

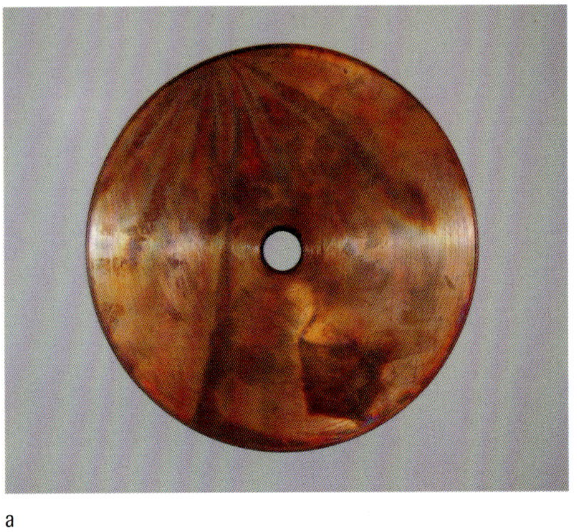

a

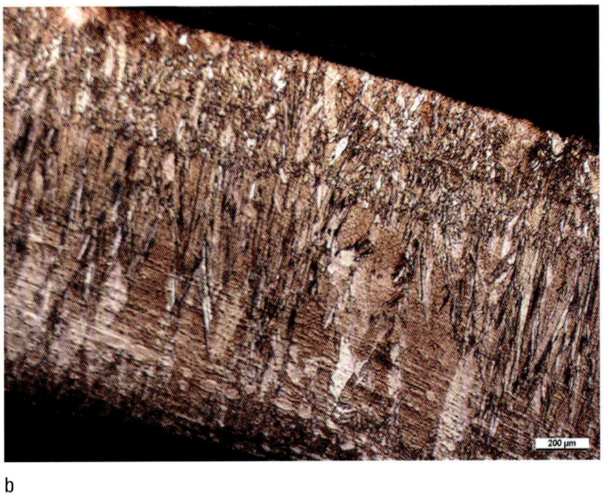

b

Figure 10.11 | A copper master used to produce LP discs (a) and a cross section showing the columnar structure typical of electrotyping (b). The differences in morphology throughout the section can be due to changes in operational variables. Photos: Centro Cultural São Paulo (a) and Virginia Costa (b).

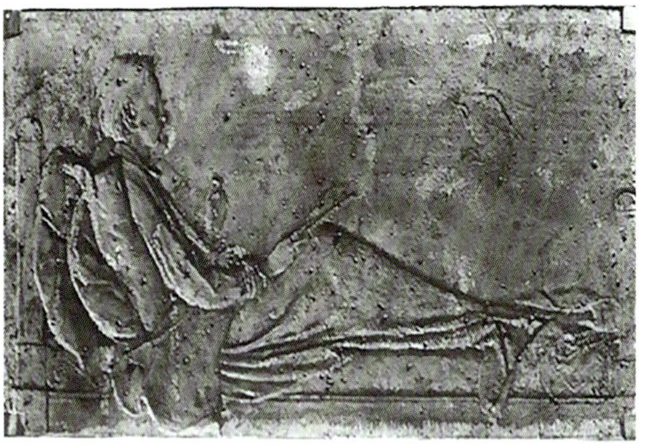

Figure 10.12 | Reverse of an electrotype showing the characteristic fine reverse details and irregular bead pattern (adapted from Beale 1975, fig. 27).

(Abramitis 2011; Scott and Stevens 2013; Metropolitan Museum of Art 2017).

Deposits obtained by electroforming are generally much thicker than electroplated ones. The process requires stringent control because it consumes much more time, and the product must be mechanically sound and have low internal stress for dimensional acceptance.

With long deposition times, high current densities at edges and surfaces closer to the anodes result in significant buildup, leading to nodules and uncontrolled growth. These nodules can often be seen on the back of electrotypes as well as in fine details (fig. 10.12) (Beale 1975).

Depending on whether the geometry of the part to be electroformed allows separation of the mold without damage to either component, permanent or expendable mandrels are used. In the 1880s, materials used for molds were plaster of paris, animal glue, gelatin, gutta-percha, and glue. Since none of these substances conducts electricity, graphite was used to coat the surfaces of the molds in order to ensure metal deposition (Scott and Stevens 2013). Today, materials of choice for expendable mandrels are 3-D printed materials, waxes, epoxies, photopolymers, and some plastics. Permanent mandrels are usually metallic (mostly aluminum, stainless steel, or titanium, because of their naturally passive surfaces) but can also be made of a conductive plastic. They can be used repeatedly until surface wear or scratching renders them useless (Canning W. and Co. 1982).

Electroforming is used in the manufacture of items such as textile printing screens, molds and dies, mesh products, bellows, compact disc stampers, radar wave guides, and optical components, which provide unique production advantages for precision operation in a number of industries. Common postplating treatments are mechanical polishing to trim surplus material from the nodular surface and machining to a regular shape (Canning W. and Co. 1982).

Electroless Plating

The term electroless plating was introduced in the 1940s to describe a method of plating metallic substrates with nickel and copper alloys without using an external power supply. Once started, the reaction proceeds continuously, producing uniform coatings with unique properties on a variety of substrates. Since the deposit catalyzes the reduction reaction, this plating process is often said to be autocatalytic.

For electroless deposition, the solution must contain two basic components: the metal species (e.g., Ni sulfate) and a reducing agent. The reducing agent supplies electrons for the reduction of the metal cations at the surface of the object. Reducing agents often include sodium hypophosphite. Besides these basic components, electroless baths contain additives, as with regular electroplating. The deposit itself is catalytic to reduction, and the reaction continues as long as the surface remains in contact with the solution (fig. 10.13) (Pearlstein 1974).

Electroless plating should not be confused with displacement plating, where a very thin film of a relatively noble metal can be deposited by simultaneously dissolving a sacrificial less noble metal. No reducing agent is required, and adherence is poor.

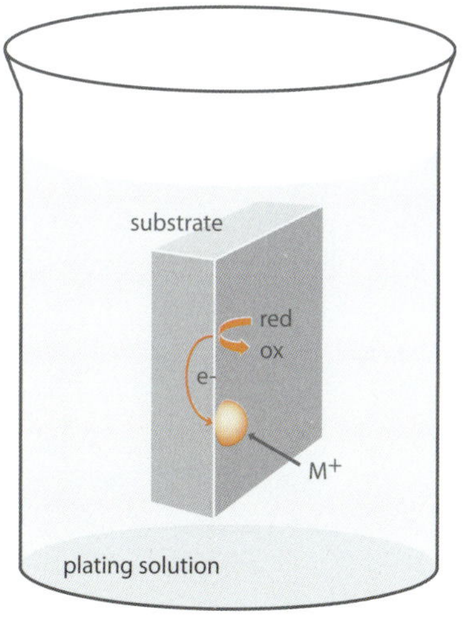

Figure 10.13 | Electroless process: The reducing agent oxidizes, supplying electrons for the reduction of metal cations at the surface of the object.

A key characteristic of electroless plating is the uniformity of the coating thickness over the entire surface in contact with the solution. Hence, grooves and blind holes have the same amount of coating as the outside of a part—in contrast to electroplating, in which the current density is not uniform when the surface presents irregularities (fig. 10.14).

An electroless nickel coating is actually an alloy with phosphorus (4%–12%), which provides natural lubricant, improving the frictional properties of the surface.

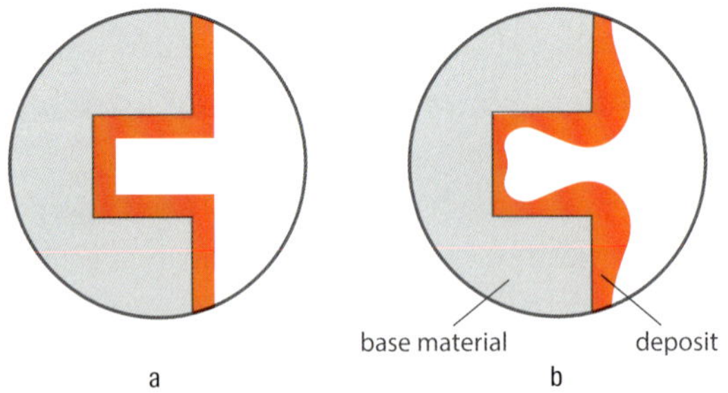

Figure 10.14 | Cross sections through components plated with nickel using electroless (a) and electrolytic (b) processes.

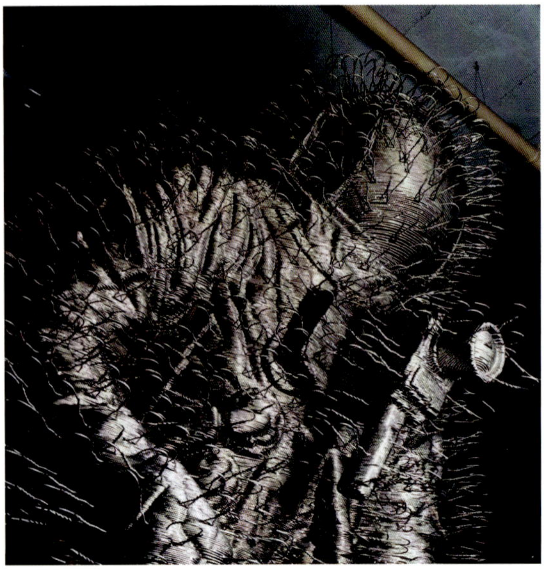

Figure 10.15 | An example of electroless nickel plating: *The Spaceman* by David Mach (2005).

Depending on the phosphorus content, such alloys may be amorphous, with improved corrosion resistance characteristics—in particular, preventing intergranular corrosion. In the as-deposited condition, the coating hardness is about 500–600 HV100. Through the use of heat treatments (200°C–400°C), deposits can be precipitation-hardened to very high levels, which produces wear resistance equal to that of commercial hard chromium coatings (1100 HV100) (Baudrand 1994).

These unique properties—excellent corrosion resistance, high wear resistance, uniformity of coating, and solderability—make electroless nickel an attractive solution for all segments of industry. Electroless copper is used especially for electronic components, from printed wiring boards to multichip modules (Paunovic 2010). Electroless nickel was used by David Mach for his 2005 sculpture *The Spaceman* (fig. 10.15).

Hot Dipping

The hot dipping process dates to the eighteenth century. It has a proven history of commercial success and can be found in almost every major application where iron and mild steel are used. Although it is a much older process than others described in this book, it is presented here so as to allow comparison and differentiation with other metallic coatings. Moreover, it is still relevant, being used to create many contemporary artifacts.

Hot dipping is a process by which a metal is coated by immersion in a bath of molten metallic salts, which progressively diffuse into the surface, forming intermetallic compound layers that strongly adhere to the substrate. Hot-dip processes are limited to metals and alloys that have a melting point low enough to allow the substrate to be pulled through the bath without tearing. Zinc, aluminum, tin, and lead coatings offer a barrier and sacrificial protection to steel; it is important that their thickness be relatively high (Wetzel 1994).

Zinc Hot Dipping (Galvanizing)

Zinc hot dipping of steel products is obtained by immersion in a bath of molten zinc. It is the most widely used of the zinc coating processes and has been commercially employed since the end of the nineteenth century. The protective coating usually consists of several layers of zinc-iron compounds produced by diffusion at high temperature during the process. This coating is much thicker than electroplated coatings. It varies from 50 to 550 μm, depending on opera-

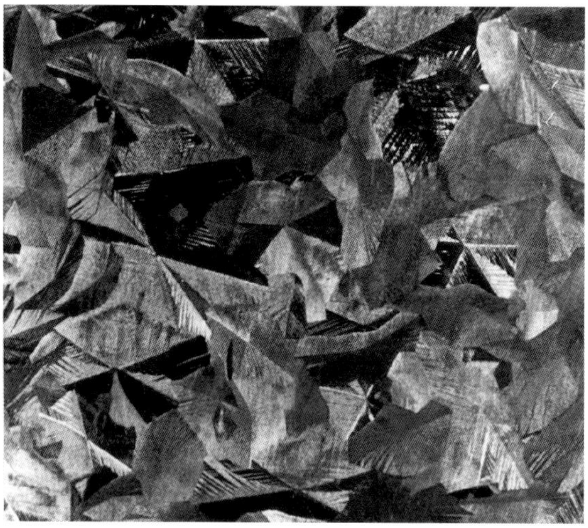

a

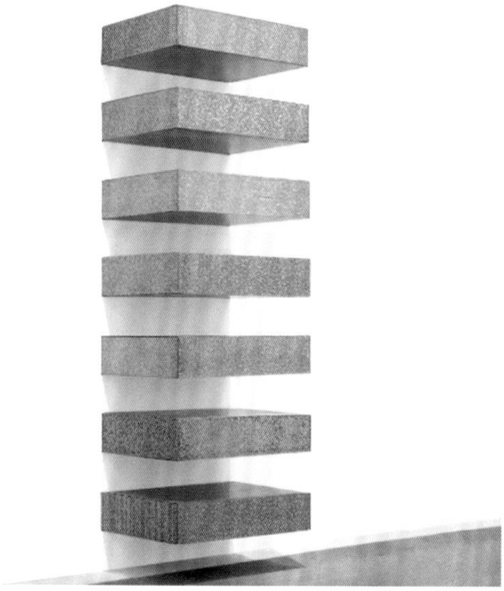

b

Figure 10.16 | Typical spangle structure of a galvanized coating (a) (reproduced from Marder 2000, fig. 31); *Untitled* by Donald Judd (1965), Moderna Museet, Stockholm (b).

tional parameters, which include duration of immersion and temperature of the bath.

For many years, galvanized articles made by hot-dip coating techniques were identified by a characteristic spangle appearance—typically a snowflake-like or sixfold star pattern visible to the naked eye (fig. 10.16). Such a well-defined spangle pattern on galvanized coatings on steel sheet was attributed to the presence of impurities—especially lead—in the bath. Because of the demand for both lead-free coatings and very smooth products, spangle size has been reduced to obtain a uniform shiny metallic appearance. More recently, spangled products may also be produced by adding small amounts of antimony (ca. 0.03%–0.10%) to the bath. Spangle size is affected by the zinc chemistry and cooling rate, as well as by other factors, such as the smoothness of the substrate (GalvInfo Center 2015b).

Zinc hot dipping can be applied in batch galvanizing (for prefabricated steel items) or on the continuous, automated lines of the steel producers (sheet, strip, wire, tubes). Coatings applied by continuous galvanizing usually contain aluminum to minimize the formation of the brittle zinc-iron alloy and to permit deep drawing and bending without damaging the coating.

Hot-dip galvanized zinc coatings have their longest life expectancy in rural areas, where sulfur dioxide and other industrial pollutant concentrations are low. These coatings also give satisfactory service in most marine environments. The protection against corrosion provided by zinc coatings is essentially determined by the thickness of the coating. Zinc coatings applied by hot dipping after fabrication are characterized by thickness (μm or mils of zinc on the surface) or by mass (g/m^2 or oz./sq. ft.) (GalvInfo Center 2015a).

Some major applications of hot-dip galvanized coatings include: electrical transmission towers and poles; bridge structural members; marine pilings and rails; highway guardrails; and galvanized and painted structural steel for aesthetic, color-coded, or extended-life applications. Wherever steel is exposed to atmospheric, soil, or water corrosion, hot-dip galvanized zinc coatings are a standard, effective, and economical method of protection.

Hot-Dip Tinning (Tin Plating)

A thin coating of molten tin is applied to the surface of iron or steel substrates, providing a nontoxic, protective, or decorative coating for food handling and packaging. The process also assists in bonding another metal to the base substrate and facilitates the soldering of a variety of components used in electronic and electrical equipment. The production of tinplate for food cans by hot dipping was replaced in the 1930s by the more efficient electrolytic process.

Lead-Alloy Hot Dipping (Terne Coating)

The term terne plate, or terne coating, historically describes a hot-dipped coating of an alloy of lead and tin on a steel substrate. Tin content is usually in the range of 2%–15%, and its presence in the molten bath forms a thin layer of a tin-iron compound, providing an adherent base for the terne coating and assisting in the formation of a relatively continuous coating.

Traditional terne metal roofing and siding materials were produced by applying lead-tin alloy over a core metal sheet of copper-bearing steel, which is usually further painted. Unpainted, it will weather to a pleasing slate-gray color. In current use, lead has been replaced by zinc, in a 50Zn-50Sn alloy. This improved version of new patented zinc-alloy coatings offers high corrosion resistance, an environmentally sensitive lead-free composition, and a low-reflective finish (ASTM International 2015c).

Aluminum Hot Dipping

The process of applying a hot-dip aluminum coating to various substrates is also referred to as *aluminizing*, in which a strong metallurgical bond forms between the coating and the steel base. The method provides a uniform coating thickness and matte surface finish. Two types of coatings are commercially available: Type 1 contains 5%–10% Si and is used in applications requiring heat and corrosion resistance; Type 2 is pure aluminum and is intended for applications requiring atmospheric corrosion resistance (ASTM International 2015a).

STRUCTURE, PROPERTIES, AND APPLICATIONS OF COATINGS

Structure

The crystal structure resulting from a plating process depends on the relative rates of formation of crystal nuclei and the growth of existing crystal and has important influence on the properties of the electrodeposit. Some of the typical structures are presented in figure 10.17.

Deposits from simple ion acidic solutions containing no addition agents usually present a columnar structure and exhibit low strength, low hardness, and high ductility. A fibrous (acicular) structure, which represents a refinement of columnar structure, may be obtained by lowering the temperature or by applying a higher current density to favor the formation of new nuclei rather than the growth of existing grains. A fine-grained structure is obtained from complex ion solutions such as cyanide or from certain additives, as well as from very high current density. With a grain size as small as 10–100 nm, such deposits are usually relatively hard and strong. A laminar (banded) deposit produces bright coatings, resulting from addition agents or deposits produced in solutions operated with either periodic reverse current or pulse plating. Laminar deposits usually have high strength and hardness but low ductility.

Hot-dip coatings, such as galvanized coating on steel, have a structure consisting of zinc-iron alloy layers metallurgically bonded to the steel base, with a relatively pure outer

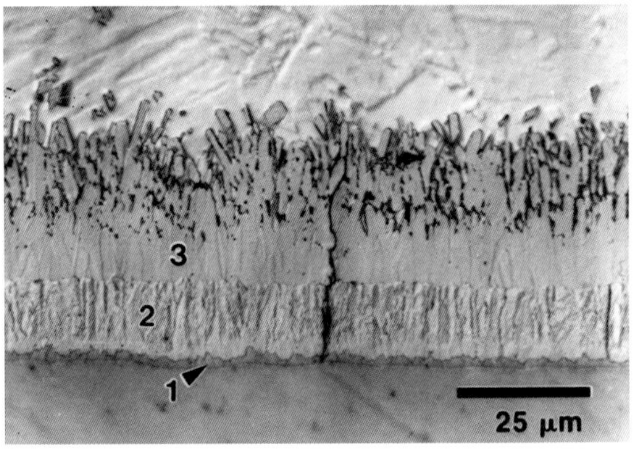

Figure 10.18 | Microstructure of Zn coating (300 s at 450°C): Starting from the steel substrate (bottom) are several intermetallic alloy layers (1, gamma phase; 2, delta phase; 3, zeta phase) and a zinc outer layer (reproduced from Marder 2000, fig. 8).

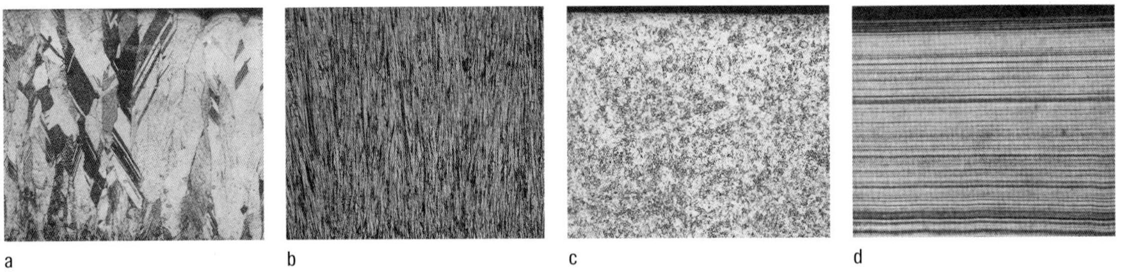

a b c d

Figure 10.17 | Cross sections showing typical structures of electroplated deposits: Columnar (a); fibrous (b); fine-grained (c); laminated or banded (d) (reproduced from Dini 1993, figs. 2–4, 6).

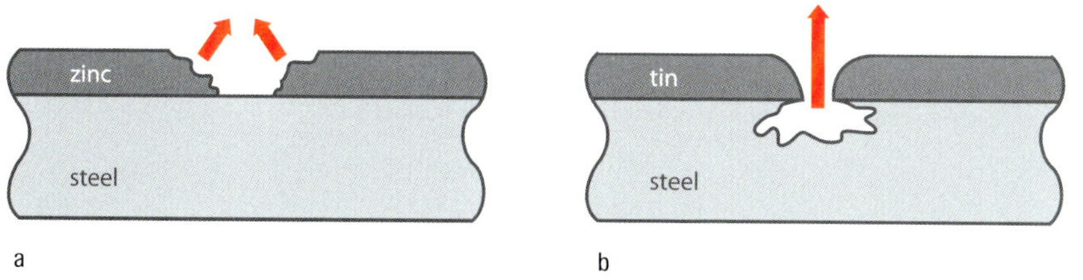

Figure 10.19 | Types of metallic coatings on steel: anodic (a); and cathodic (b).

zinc layer (fig. 10.18). The complex structure of layers in a galvanized coating varies greatly in chemical composition and physical and mechanical properties.

Properties

Among the many applications of electrodeposits, the most important one is corrosion protection of the base metal. With regard to corrosion protection characteristics, metallic coatings fall into two major groups, those that are anodic (electronegative) to the base metal and those that are cathodic (electropositive) (Safranek 1974, 1–24).

In the first group, protection is provided by a galvanic mechanism. The coating is in fact less noble than the underlying substrate and therefore corrodes preferentially, which confers galvanic protection to any substrate area that may be exposed at pores or cut edges. Protection is effective as long as the coating has sufficient mass to be sacrificially consumed (fig. 10.19a).

In the cathodic type, protection relies on a barrier mechanism. The coating is a physical barrier that isolates the underlying metal from the corrosive environment. Barrier coatings should therefore corrode at a slower rate than the substrate. A continuous envelope around the article is mandatory, since corrosion of the base metal due to a galvanic attack would be accelerated at any discontinuities in the coating (fig. 10.19b).

In general, zinc and cadmium coatings constitute the principal electrodeposited finishes of the anodic type on steel, whereas all the other commercially electrodeposited metals fall into the cathodic group. It is important to remember that the potential difference that exists between coating and base metal depends not only on the two metals but also on the environment to which they are exposed. Therefore, the standard electromotive force series (see fig. 2.5) provides only a very rough indication of the difference of potential that may exist between them under actual conditions. An overview of important characteristics of selected plated coatings is presented in table 10.2.

Testing methods have been developed over the years to ensure that metallic coatings have the expected properties

Table 10.2. | Properties of selected plated coatings.

	Coatings											
Properties	Cd	Cr	Cu	Cu alloys	Au and Au alloys	Pb and Pb-Sn	Ni (electrolytic)	Ni (electroless)	Pt group metals	Ag	Sn and Sn alloys	Zn
Abrasion resistance		X						X	X			
Conductivity			X	X	X					X		
Corrosion resistance	X	X			X	X	X	X	X		X	X
Decorative appearance		X		X	X		X	X	X	X	X	
Contact resistance					X				X			
Non–stick resistance		X										
Softness and lubricity	X					X				X	X	
Solderability	X		X	X	X	X	X	X			X	
Rebuilding worn parts		X	X				X					

Source: Data from Edwards 1983, introduction.

for a given application. These tests are mostly used as part of the production control process or in the development of new coating applications. They are often destructive, and some require a considerable amount of sampling, which is not acceptable in the case of cultural heritage. Nevertheless, it is useful to know that the common properties of metallic coatings may be evaluated and that standards exist for determining the thickness of deposit, wear resistance, reflectivity, contact resistance, porosity, tensile strength, electrical conductivity, adhesion, internal stress, corrosion resistance, hardness, solderability, and magnetic properties (Safranek 1974; Ogburn 1974; Simpson and Townsend 2005).

Applications of Common Coatings

Some common metallic coatings are briefly described below (Lowenheim 1974).

Cadmium

The main purpose of Cd plating is to provide an attractive, corrosion-protective coating over various base metals, especially iron and steel. Cadmium is anodic to iron, protecting the underlying ferrous metal at its own expense, even when it becomes scratched and exposes the substrate. For parts used indoors or in sheltered positions outdoors, thin coatings (<25 μm) are used, because of the relatively high price of Cd. Aircraft, marine, and military outdoor uses of thicker coatings are common. The electrical industry makes use of Cd plate on steel and other metals because it is easily soldered and has low contact resistance. However, the toxicity of cadmium discourages its use. Common postplating treatments are: chromate conversion coating, lacquering and painting, and hydrogen embrittlement relief (by heat treatment).

Chromium

Chromium coatings exhibit high corrosion resistance upon atmospheric exposure. There are two principal classes of chromium deposits: decorative chromium, where thin coatings serve as nontarnishing, durable surface finishes; and *hard chromium*, or industrial chromium, where heavy coatings are used to take advantage of the special properties of chromium, including resistance to heat, wear, corrosion, and erosion, as well as a low coefficient of friction. Postplating treatments are usually not necessary because of the great passivity and tarnish resistance of Cr.

Decorative chromium coatings are very thin—usually between 0.13 and 1.3 μm—and have no protective value when applied directly to steel or brass. Therefore, they are applied over a nickel or copper-nickel undercoating and generally reproduce the finish of the substrate. The chromium layer is often deliberately deposited in a crazed, or microcracked, condition in order to enhance corrosion resistance by minimizing adverse galvanic effects between the cathodic chromium and the anodic nickel (fig. 10.20).

Hard chromium coatings are thick (2.5–50 μm; exceptionally, up to 250 μm). This kind of coating is primarily intended to increase the service life of functional parts by enhancing their resistance to wear, abrasion, heat, and corrosion; deposits are also applied to restore the dimensions of undersized parts.

Cobalt

The magnetic properties of cobalt and its alloys are of interest in electronic applications, such as memory drums, disks, cards, and tapes, particularly in the computer industry. Cobalt alloys have been investigated for electroforming parts for the aerospace industry because of their low coefficient of friction and excellent wear resistance.

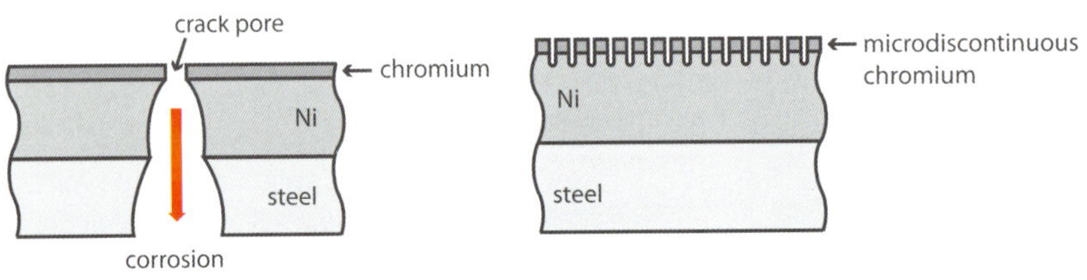

Figure 10.20 | Corrosion of chromium coatings: For regular chromium, local defects allow the corrosion to develop in depth (a); for microdiscontinuous chromium, because the discontinuities are small and regularly distributed over the surface, the galvanic effect is neutralized (b) (adapted from Nickel Institute 2014, fig. 10).

Copper

Copper coatings are rarely used alone in applications where a durable, attractive surface is required because they tend to tarnish and stain rapidly when exposed to the atmosphere. They are widely used, however, in combination with nickel and chromium coatings. Cu electrodeposits are used widely as underplates in multiple-plate systems, as stop-offs for heat transfer, and in printed circuit boards.

Gold

Used initially mostly as a decorative finish, gold became a functional utilitarian coating with the development of electronics in the 1950s. There are hundreds of formulations for electroplating or electroless gold plating baths. Decorative gold plating is mostly flash-plated over a bright substrate consisting of nickel-plated brass or other base metal such as stainless steel in order to achieve a desired color. Alloying changes the color: Cu makes the deposit pinker; Ni makes it paler or whiter; Ag and Cd make it greener. Pure gold is preferred for its electrical conductivity, corrosion resistance, and solderability.

Iron

Iron is mainly used for applications that depend on the desirable physical properties of iron and its low cost. Early applications included the production of electrotypes and the building up of worn or underdimensioned machine parts. The good magnetic properties that can be developed in electrolytic iron led to its use in the cores of induction coils before the development of Fe-Ni alloys with even better magnetic properties. Iron plating was used during World War II to make electrotypes and coat stereotypes (long-play matrix), in order to limit the wear of the base materials, such as Ni and Cu.

Lead

Lead has not been extensively deposited by electroplating because its low melting point of 325°C facilitates application by hot dipping. Electroplated lead has been used for corrosion protection of electrical fuse boxes installed in industrial plants or where sulfur-bearing atmospheres are present. Lead is also codeposited with tin for wire plating, automotive crankshaft bearings, and printed circuits.

Nickel

Ni plating is one of the oldest protective-decorative electrodeposited metallic coatings for steel, brass, and other metallic substrates. More recently it has been applied to plastics of various kinds. Electrodeposits of nickel possess a wide variety of properties, depending on the composition of the plating bath and on operating conditions. They can be classified according to application or appearance as:

- *General purpose*: Used primarily to protect iron, copper, or zinc alloys against corrosive attack in rural, marine, and industrial atmospheres, with recommended thicknesses varying from 8 to 50 μm. Heavy deposits are used to build up worn or undersized parts and to provide protection against corrosive chemical environments.
- *Bright nickel*: Used to provide decorative finishes on metals that tarnish without suitable protection. Since they have sulfur-containing brightening agents, their corrosion resistance may be reduced. They are often applied as undercoatings for chromium or one of the precious metals or as overcoatings for a sulfur-free nickel deposit (a combination known as *duplex nickel*) (fig. 10.21).
- *Black nickel*: Has little protective value and is used primarily to obtain a dark, nonreflective decorative finish.

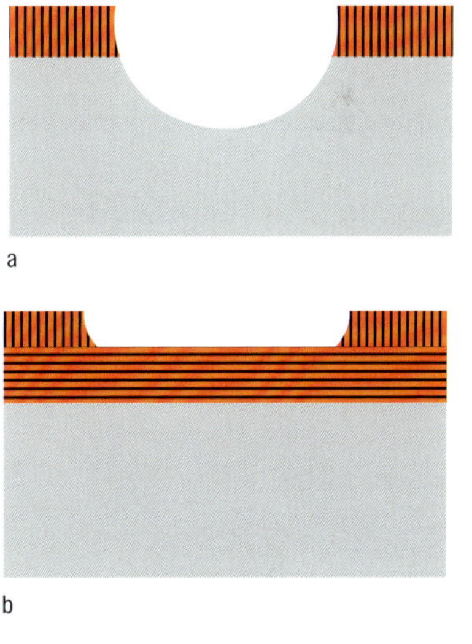

Figure 10.21 | Behavior of nickel coatings under corrosive conditions: A single-layer bright nickel deposit with deep corrosion penetration (a); a duplex deposit consisting of a lower layer of semibright and a top layer of bright nickel (b), in which the low-sulfur semibright deposit acts as a barrier to further penetration once the top layer has been disrupted, resulting in a lateral spread of corrosion (adapted from Canning W. and Co. 1982, 395).

Rhodium

Rhodium is used to provide a whiter finish on platinum goods and as a nontarnishing coating on sterling silver products. As a nontarnishing coating, however, rhodium can accelerate the corrosion of the base metal by galvanic couple effect at the inevitable pinholes within the layer. In addition, two properties of Rh make it suitable as a coating in many electronic applications: low electrical resistivity and high hardness.

Silver

Used primarily for its decorative appeal on tableware and jewelry, silver coatings also have industrial applications because of silver's versatile properties. Silver coatings are used for their high conductivity and low contact resistance in the electrical industry. Moreover, silver is an excellent heavy-duty sleeve bearing material and is used extensively for rotating parts in the aircraft power plant industry. Because silver forms the characteristic brown to black silver sulfide film when exposed to sulfur-containing gases, some objects are lacquered or passivated. Certain passivating treatments are effective in inhibiting sulfide tarnish, but these films are thin and fragile and will not withstand abrasion.

Tin

Electrodeposited tin coatings provide excellent solderability, ductility, and softness, and they impart corrosion or tarnish resistance to stronger substrate materials, which are required for their engineering properties. The largest use of tin electrodeposits is in the production of tinplate at steel mills, primarily for use in food preservation containers. A thin tin coating (typically 0.4–1.6 μm) protects the steel inside a can as long as the environment is oxygen-free. The second-largest use is in the electronics industry, where many surfaces requiring good solderability and corrosion or tarnish resistance are coated.

Zinc

Because it is relatively inexpensive, zinc is often preferred for coating iron and steel parts when protection from either atmosphere or indoor corrosion is the primary objective. Pure zinc coatings are usually applied using the hot-dip or electroplating method. The hot-dip process produces a thick coating (300–900 g/m^2) containing an intermetallic layer at the zinc-steel interface (see fig. 10.18). Electroplated zinc coatings are thinner (up to 100 g/m^2) and more uniform, and they have a bright, silvery appearance. Since they become dull gray after exposure to air without subsequent treatment, nearly all plated zinc is followed by some type of chromate dip or a coating of clear lacquer. These preserve the appearance of the part and vastly increase the bright shelf life of the surface. Bright zinc deposits are used for a wide variety of low-cost consumer goods. Zinc coatings offer better corrosion protection than cadmium coatings in industrial environments, but cadmium retains the initial appearance longer and develops fewer corrosion products in mild environments.

CHARACTERIZATION

It is difficult to determine visually the presence of metallic coatings on artifacts, especially if they are undamaged. Some indication can be obtained by the positive response to a magnet, indicating the presence of a coated iron-alloy substrate despite a golden shade or a bright, unrusted silvery surface.

Once its presence is confirmed, identification of the plating layer can be done from its color. A red or yellow aspect indicates copper or gold (or alloys of these elements). Silvery surfaces offer more possibilities, as shown in table 10.3.

Additional information may be given by the category of the artifact (see table 10.1)—for example: use (functional, decorative, or both), cost (expensive and unique could justify a noble coating), environment (Zn coatings finished with chromate may be used indoors, but Ni coatings would not be chosen for outdoor use without a final, very thin layer of chromium to avoid tarnishing), and shape (chromium has poor throwing power and therefore would not be used for artifacts with complex geometries).

Table 10.3. | Identification of silvery electrodeposits.

Coating color	Metal present
Silvery white	Cd
Silvery blue	Zn
Silvery with bluish tinge	Cr
Silvery with yellow tinge	Ni
Silvery, highly reflective, tarnishes	Ag
Silvery with white tinge, turns gray slowly	Sn

Source: Data from Child 1993a, 50.

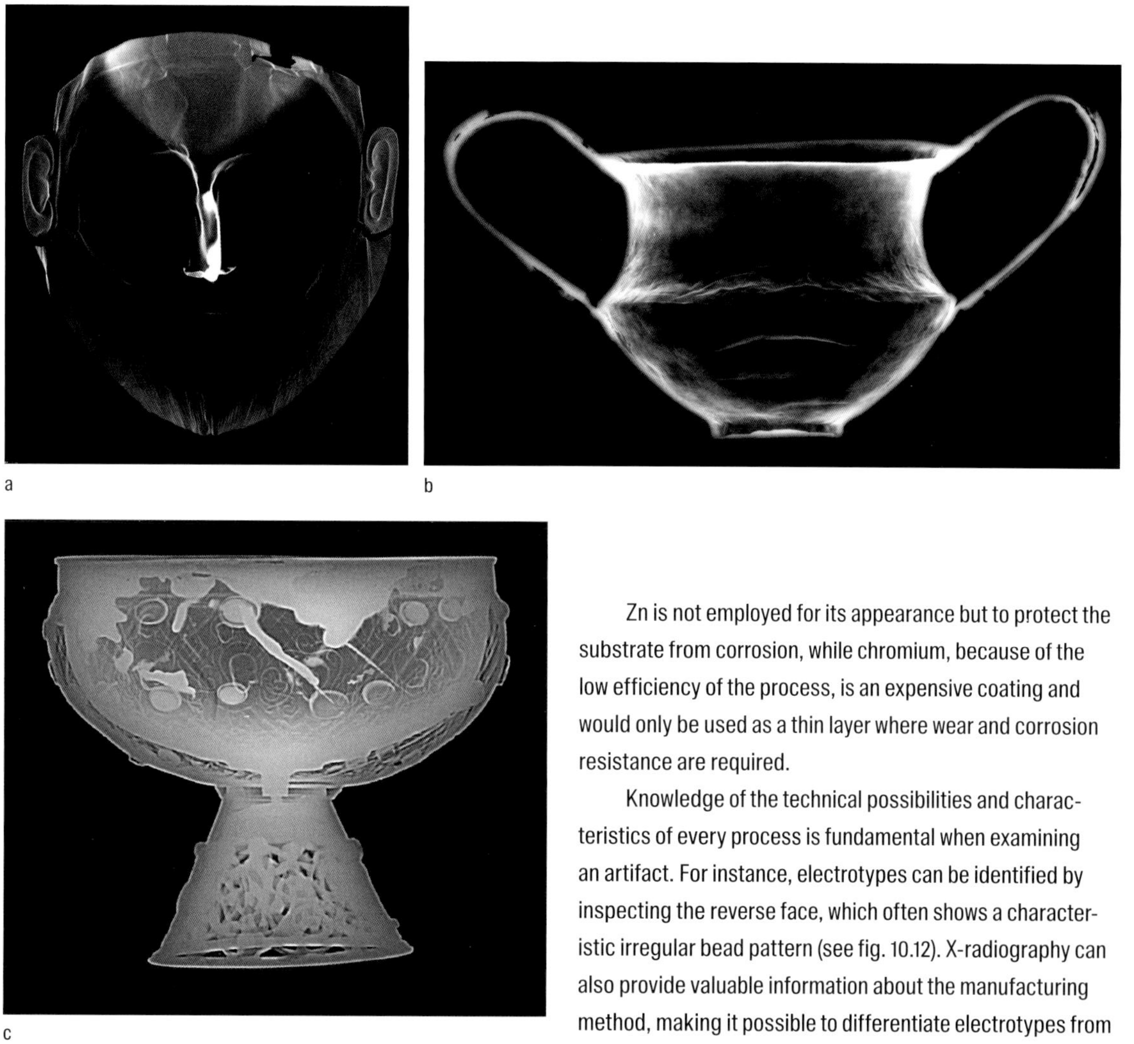

a b c

Figure 10.22 | Radiograph of the replica of the *Mask of Agamemnon* from Mycenae, exhibiting the texture remaining from the electroforming process (a); radiograph of a Late Helladic (ca. 1550–1500 BCE) gold kantharos exhibiting hammered texture from raising (b) (reproduced from Abramitis 2011. On file in the Sherman Fairchild Center for Objects Conservation, the Metropolitan Museum of Art, New York. Emile Gilliéron; Maker: Würtembergische Metallwarenfabrik. Dodge Fund, 1906 [06.224]); Radiograph of the Maître Alpais electrotype ciborium, where the inner lining has been soldered in place (c) (reproduced from McLeod, Campbell, and Nouvel 2010).

Zn is not employed for its appearance but to protect the substrate from corrosion, while chromium, because of the low efficiency of the process, is an expensive coating and would only be used as a thin layer where wear and corrosion resistance are required.

Knowledge of the technical possibilities and characteristics of every process is fundamental when examining an artifact. For instance, electrotypes can be identified by inspecting the reverse face, which often shows a characteristic irregular bead pattern (see fig. 10.12). X-radiography can also provide valuable information about the manufacturing method, making it possible to differentiate electrotypes from worked or cast artifacts (fig. 10.22).

Complementary analytical characterizations, such as X-ray fluorescence (XRF), should be applied to check starting hypotheses. The scientist in charge of the analysis must have enough experience with metallic coatings to correctly interpret the results and distinguish among the different layers—not simply record them as an average composition, as if the detected elements were part of an alloy (see p. 42). The important difference lies not in the detected elements but in their distribution in the material.

Unfortunately, there is no nondestructive technique to disclose the presence of undamaged, intact surface coatings. Sampling is necessary. A few cubic millimeters of the artifact may be sufficient to prepare metallographic cross sections and examine them under an optical microscope. Complementary scanning electron microscopy/energy-dispersive X-ray spectroscopy (SEM/EDS) observations and microhardness measurements provide a very complete picture of the internal structure and the coating and the chemical composition of each part (fig. 10.23).

Thorough knowledge of the different coating technologies is invaluable. For example, when determining whether a Ni coating was applied by electroplating or electroless plating, recalling that the electroless process includes phosphorus in the alloy's composition may lead to using an analytical technique that can detect phosphorus.

Regarding condition assessment, the presence of defects on metallic coatings affects the aesthetic appearance of the part as well as its corrosion resistance. Examination of defects such as dullness, porosity, and blistering may be helpful to determine the presence of coatings and hypothesize about their nature. Defects may originate from the object's manufacture, because of inappropriate substrate preparation or problems during plating. They may also have been created during the object's life, because metallic coatings are easier to damage than bulk materials owing to their limited thickness. A list of familiar defects and information about their likely causes may be found in the literature (Morse 1984; Galvanizeit 2017).

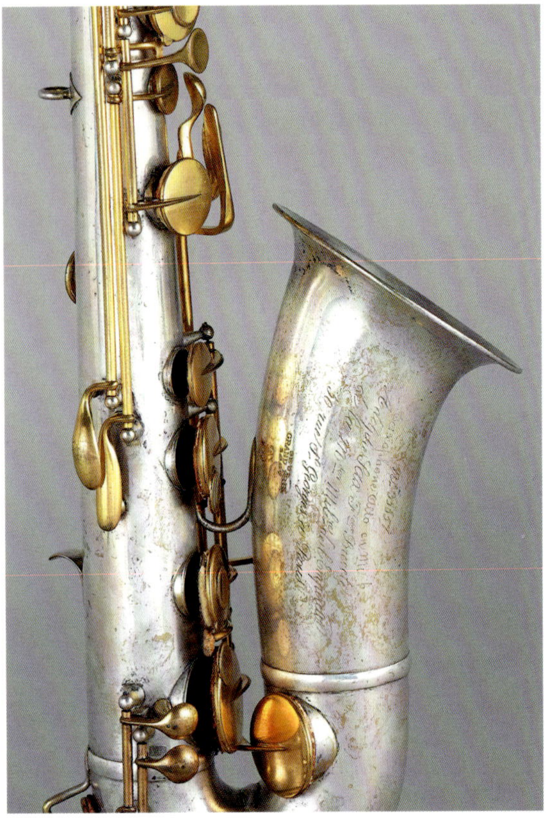

a

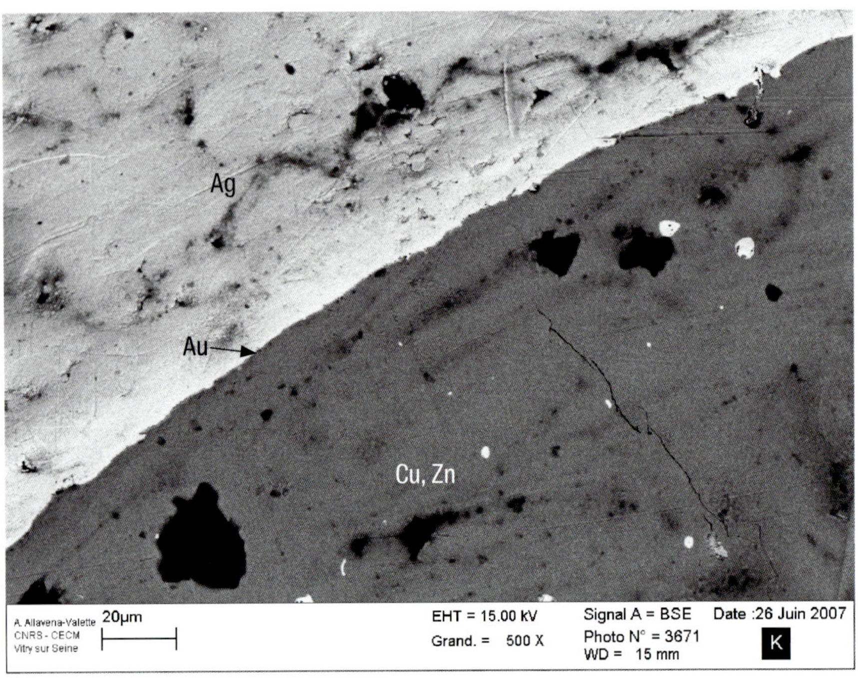

b

Figure 10.23 | Alto saxophone by Adolphe Sax (1871), Musée de la Musique, Paris (a); SEM of a polished extremity of the saxophone, consisting of a leaded Cu-Zn alloy, coated with a very thin Au layer and an external, thicker Ag layer (b). Photo: J. Marc Anglès (a); image: Virginia Costa (b).

CONSERVATION-RELATED LITERATURE

Publications on electroplating in the conservation field cover a diversity of artifacts. In most cases, they are about the investigation and characterization of objects manufactured using this technique, such as Marcel Duchamp's electroplated plasters (Lins 2009), Richard Lippold's gold-plated wire sculpture (Flotte and Dommermuth 2000), early twentieth-century cadmium-plated instruments (Scott and Derrick, 2007), modern metallic threads (Járó, Gál, and Tóth 2000), Goya's engraving plates (Bautista, Lage, and Velasco 2007), and tin cans in museum collections (Brambilla, Michel, and Bertholon 2016; Caiazzo, Brambilla, and Mischler 2016). There are also references to the use of electroplating to restore deteriorated surfaces (Warren-Findley 1971; Ogburn et al. 1972), as well as a short review paper on electroplated coatings (Lewis 1988). The use of adhesives to consolidate nickel-chrome coatings that were peeling off has been systematically investigated in a MSc dissertation (Rouquié 2008).

Regarding electroforming, some papers—especially German publications—address historical, technical, and restoration-related issues (Meissner and Doktor 2000; Haber and Heimler 1994). Others focus on describing materials and technical features—information that can assist in identifying techniques (Scott and Stevens 2013; Abramitis 2011; Wharton 1984; Clare and Lins 2017)—and some report on restoration procedures adopted in various cases (Ferreira 1998; Doktor, Meissner, and Mach 2000; Haber and Heimler 1994).

As for galvanized artifacts (hot dip), there are publications in the conservation field dealing with their characterization (Nagy and Adamsons 2007; Fryer, Pullen, and Greenfield 2010), treatment (Day 1997; Ryan 1995), and repair or replacement (Shiffer 1991; Pieper 1990).

REFERENCES AND BIBLIOGRAPHY

Abramitis, D. H. 2011. "The Mask of Agamemnon: An Example of Electroformed Reproduction of Artworks Made by E. Gilliéron in the Early Twentieth Century." In *Now at the Met Blog. https://www.metmuseum.org/blogs/now-at-the-met/features/2011/mask-of-agamemnon.*

ASTM International. 2015a. *ASTM A463/A463M. Standard Specification for Steel Sheet, Aluminum-Coated by the Hot-Dip Process*. West Conshohocken, PA: ASTM International.

——. 2015b. *ASTM A780/A780M. Standard Practice for Repair of Damaged and Uncoated Areas of Hot-Dip Galvanized Coatings*. West Conshohocken, PA: ASTM International.

——. 2015c. *ASTM A1074. Standard Specification for Hot Tin and Hot Tin/Lead Dip on Ferrous and Non-Ferrous Metals*. West Conshohocken, PA: ASTM International.

Baudrand, W. 1994. "Properties of Electroless Nickel-Phosphorus Coatings." In *Surface Engineering*, edited by ASM Handbook Committee, 290–309. Vol. 5 of *ASM Handbook*. Materials Park, OH: ASM International.

Bautista, A., M. Lage, and F. Velasco. 2007. "Electrochemical Study of the Corrosion Behavior of Electroplated Copper Plates." In *Strategies for Saving our Cultural Heritage: Papers Presented at the International Conference on Conservation Strategies for Saving Indoor Metallic Collections with a Special Section on Legal Issues in the Conservation of Cultural Heritage, Cairo, 26–28 February*, edited by V. Argyropoulos, A. Hein, and M. A. Harith, 110–14. Athens: Department of Conservation of Antiquities and Works of Art, Technological Educational Institute of Athens.

Beale, A. 1975. "A Technical View of Nineteenth-Century Sculpture." In *Metamorphoses in Nineteenth-Century Sculpture*, edited by J. L. Wasserman, 47. Cambridge, MA: Harvard University Press.

Brambilla, L., A. Michel, and R. Bertholon. 2016. "Condition of Cans in Collections: A Challenge in Conservation." In *Metal 2016: Proceedings of the Interim Meeting of the ICOM-CC Metals Working Group, September 26–30, New Delhi, India*, edited by R. Menon, C. Chemello, and A. Pandya, 266–74. Paris: ICOM-CC.

Brenner, A. 1963. *Electrodeposition of Alloys: Principles and Practice.* Vol. 1. New York: Academic Press.

Browning, M. E. 1994. "Selective (Brush) Plating." In *Surface Engineering*, edited by ASM Handbook Committee, 277–81. Vol. 5 of *ASM Handbook*. Materials Park, OH: ASM International.

Caiazzo, F. C., L. Brambilla, and S. Mischler. 2016. "Preserving Cans in Collections: The Importance of Modeling Corrosion of Tin Cans for Cultural Heritage." In *Metal 2016: Proceedings of the Interim Meeting of the ICOM-CC Metals Working Group, September 26–30, New Delhi, India*, edited by R. Menon, C. Chemello, and A. Pandya, 332. Paris: ICOM-CC.

Canning W. and Co. 1982. "Electroforming." In *The Canning Handbook: Surface Finishing Technology, Integrated Design*. 23rd ed., 573–90. Birmingham: Kluwer Academic Publishers.

Child, R. 1993a. *Metal Plating and Patination*, edited by S. LaNiece. Oxford: Butterworth-Heinemann.

——. 1993b. "Modern Electroplating and Electrofinishing Techniques." In *Metal Plating and Patination*, edited by S. LaNiece, 291–300. Oxford: Butterworth-Heinemann.

Clare, T. L., and A. Lins. 2017. *Finishing Techniques in Metalwork: Electroforming.* Philadelphia: Philadelphia Museum of Art. *http://www.philamuseum.org/booklets/7_42_74_1.html.*

Clauss, R. J. 1974. "Selection of Coatings." In *Modern Electroplating*, 3rd ed., edited by F. A. Lowenheim, 656–72. New York: Wiley InterScience.

Day, C. 1997. "A Technical Note on the Treatment of Tin Plate: The Five Percent Solution." *Journal of the Canadian Association for Conservation* 22:77–80.

Dini, J. W. 1993. *Electrodeposition: The Materials Science of Coatings and Substrates*. Westwood, NJ: Noyes Publications.

Doktor, A., B. Meissner, and M. Mach. 2000. "Galvanoplastik—chemische Analysen und Restaurierung." In *Bronze- und Galvanoplastik: Geschichte–Materialanalyse–Restaurierung*, edited by B. Meissner, A. Doktor, and M. Mach, 138–51. Lampertswalde, Germany: Stoba-Druck.

Edwards, J. 1983. *Electroplating: A Guide for Designers and Engineers.* Birmingham: Institute of Metal Finishing, Committee for the Promotion of Electroplating,

Faust, C. F. 1974. "Principles of Alloy Plating." In *Modern Electroplating*, 3rd ed., edited by F. A. Lowenheim, 486–505. New York: Wiley InterScience.

Ferreira, J. 1998. "Histoire de la Galvanoplastie. Étude et Conservation de Deux Sculptures: Nu Féminin portant un Vieillard sur les Genoux par A. Rodin et Fontaine à la Grenouille par Houguenade. Étude Technico-scientifique: Réconstitution par Galvanoplastie d'un Élément manquant de la Fontaine à la Grenouille." Mémoire de fin d'études [diss.], Institut National du Patrimoine.

Flandro, X. C., and K. Loveman. 2016. "It's Electric! Manufacturing Techniques and Architectural Uses of Galvanoplasty and Electroplating." In *Metal 2016: Proceedings of the Interim Meeting of the ICOM-CC Metals Working Group, September 26–30, New Delhi, India*, edited by R. Menon, C. Chemello, and A. Pandya, 76–82. Paris: ICOM-CC.

Flotte, J., and J. Dommermuth. 2000. "Richard Lippold's The Sun: Investigation and Treatment Considerations." In *Gilded Metals: History, Technology and Conservation,* edited by T. Drayman-Weiser, 329–36. London: Archetype Publications.

Fryer, E., D. Pullen, and D. Greenfield. 2010. "Saving your Spangles: The Conservation and Care of Galvanized Steel Sculptures." In *Metal 2010: Proceedings of the Interim Meeting of the ICOM-CC Metal Working Group, October 11–15, Charleston, South Carolina*, edited by P. Markidian, C. Chemello, C. Watters, and P. Hull, 261–68. Clemson, SC: Clemson University.

Galvanizeit. 2017. "Types of Inspection: Finish and Appearance." *https://galvanizeit.org/inspection-course/types-of-inspection/finish-appearance/finish-visual-defects.*

GalvInfo Center. 2015a. "Selecting Coating Thickness (Weight or Mass) for Galvanized Steel Sheet Products." GalvInfoNote 1.6. Durham, NC: International Zinc Association. *https://galvinfo.zinc.org/wp-content/uploads/sites/8/2017/05/GalvInfoNote_1_6.pdf.*

———. 2015b. "The Spangle on Hot-Dip Galvanized Steel Sheet." GalvInfoNote 2.6. Durham, NC: International Zinc Association. *https://galvinfo.zinc.org/wp-content/uploads/sites/8/2017/05/GalvInfoNote_2_6.pdf.*

Geissman, W. C., and D. J. Bennet. 1974. "Brass." In *Modern Electroplating*, 3rd ed., edited by F. A. Lowenheim, 506–27. New York: Wiley InterScience.

Haber, G. J., and M. Heimler. 1994. "Kupfergalvanoplastik: Geschichte, Herstellungstechniken und Restaurierungsproblematik kunstindustrieller Katalogware." In *Metallrestaurierung: Beiträge zur Analyse, Konzeption und Technologie*, edited by P. Heinrich, 160–81. Munich: Georg D. W. Callwey.

Izaki, M. 2000. "Electrodeposition of Iron and Iron Alloys." In *Modern Electroplating*, 4th ed., edited by M. Schlesinger and M. Paunovic, 461–82. New York: Wiley InterScience.

Járó, M., T. Gál, and A. Tóth. 2000. "The Characterization and Deterioration of Modern Metallic Threads." *Studies in Conservation* 45 (2): 95–105.

Kuzmic, J. J. 1990. "Plating on Plastics." In *Electroless Plating: Fundamentals and Applications*, edited by G. O. Mallory and J. B. Hajdu. Orlando, FL: American Electroplaters and Surface Finishers Society.

Lewis, L. 1988. "Properties and Applications of Electro-Plated Coatings." In *Modern Metals in Museums*, edited by R. E. Child and J. M. Townsend, 39–40. London: Institute of Archaeology.

Lins, A. 2009. "Marcel Duchamp's Electroplated Plasters at the Philadelphia Museum of Art." In *Marcel Duchamp—Étant donnés*, edited by M. R. Taylor, 230–39. Philadelphia: Philadelphia Museum of Art.

Lowenheim, F. A. 1974. *Modern Electroplating.* New York: John Wiley and Sons.

Marder, A. R. 2000. "The Metallurgy of Zinc-Coated Steel." *Progress in Materials Science* 45:191–271.

McLeod, B., M. Campbell, and O. Nouvel. 2010. "A Christofle Electrotype of the Medieval Maître Alpais Ciborium." In *The Heritage of "Maître Alpais": An International and Interdisciplinary Examination of Medieval Limoges Enamel and Associated Objects*, edited by S. La Niece, S. Röhrs, and B. McLeod, 33–37. London, British Museum.

Meissner, B., and A. Doktor. 2000. "Galvanoplastik—Geschichte einer Technik aus dem 19. Jahrhundert." In *Bronze- und Galvanoplastik: Geschichte–Materialanalyse–Restaurierung*, edited by B. Meissner, A. Doktor, and M. Mach, 127–37. Lampertswalde, Germany: Stoba-Druck.

Metropolitan Museum of Art. 2017. *An Art of Attraction: The Electrotyping Process*. Informational video. New York: Metropolitan Museum of Art. *https://www.youtube.com/watch?v=iTytvWs5nV8.*

Morse, L. M. 1984. "Testing Electrodeposit Coatings: Inspection." In *Electroplating Engineering Handbook*, 4th ed., edited by L. J. Durney, 332–40. New York: Van Nostrand Reinhold.

Nagy, E., and K. Adamsons. 2007. "Saving Judd's 'Untitled 1964': Revival of a Galvanized Steel Single Stack Sculpture with Red Nitrocellulose Paint." *Studies in Conservation* 46 (3): 245–61.

Nickel Institute. 2014. *Nickel Plating Handbook*. Brussels: Nickel Institute. *https://www.nickelinstitute.org/~/media/Files/TechnicalLiterature/NPH_141015.ashx.*

Ogburn, F. 1974. "Methods of Testing." In *Modern Electroplating*, 3rd ed., edited by F. A. Lowenheim, 673–709. New York: Wiley InterScience.

Ogburn, F., E. Passaglia, H. C. Burnett, J. Kruger, and M. L. Picklesimer. 1972. "Structural Repair and Surface Restoration of Four Twentieth Century Gilded Bronze Equestrian Statues." In *Preprints of the ICOM-CC, Madrid,* vol. 2, 2–13. Paris: International Council of Museums.

Parkinson, R., and T. Hart. 1995. "Electroplating on Plastics." NiDI Technical Series no. 10078. Toronto: Nickel Development Institute.

Paunovic, M. 2010. "Electroless Deposition of Copper." In *Modern Electroplating*, 5th ed., edited by M. Schlesinger and M. Paunovic, 433–46. Hoboken, NJ: John Wiley and Sons.

Pavlova, O. I. 1963. *Electrodeposition of Metals: A Historical Survey.* Jerusalem: S. Monson.

Pearlstein, F. 1974. "Electroless Plating." In *Modern Electroplating*, 3rd ed., edited by F. A. Lowenheim, 486–505. New York: Wiley InterScience.

Pieper, R. 1990. *Restoring Metal Roof Cornices*. Preservation Tech Notes, Metals, no. 2. Washington, DC: National Park Service.

Raub, C. 1993. "The History of Electroplating." In *Metal Plating and Patination,* edited by S. La Niece, 284–90. Oxford: Butterworth-Heinemann.

Rouquié, E. 2008. "Nickel-chrome. Conservation-restauration de mobilier métallique à revêtement électrolytique: une table basse de Félix Aublet (1932) et un porte-serviettes de Colette Guéden (1946). Recherche d'un adhésif de refixage du nickelage soulevé." MSc diss., Institut National du Patrimoine.

Ryan, R. 1995. "Tin Plate and Its Treatment." In *Where to Start, Where to Stop? Papers from the British Museum/MEG Ethnographic Conservation Colloquium, In Memory of Harold Gowers*, edited by L. Hill and S. Giles, 145–50. Museum Ethnographers Group Occasional Papers, no. 4. Hull, UK: Museum Ethnographers Group.

Safranek, W. H. 1974. *The Properties of Electrodeposited Metals and Alloys: A Handbook*. Orlando, FL: American Electroplaters and Surface Finishers Society.

Scott, D. A., and M. R. Derrick. 2007. "Deterioration of Cadmium-Coated Instruments in Museum Storage." *Studies in Conservation* 52 (1): 59–68.

Scott, D. A., and D. Stevens. 2013. "Electrotypes in Science and Art." *Studies in Conservation* 58 (3): 189–98.

Shiffer, R. A. 1991. *In-Kind Replacement of Historic Stamped Metal Exterior Siding*. Preservation Tech Notes, Metals, no. 3. Washington, DC: National Park Service.

Sifco. 2017. *www.sifcoasc.com*.

Simpson, T. C., and H. E. Townsend. 2005. "Metallic Coatings." In *Corrosion Tests and Standards: Application and Interpretation*. 2nd ed., edited by R. Baboian, 517–24. West Conshohocken, PA: ASTM International.

Smith, A. 1978. "A History of Tinplate." *Tin International* 51 (3): 88–91.

Strow, H. 1994. "Copper Alloy Plating." In *Surface Engineering*, 255–57. Vol. 5 of *ASM Handbook*. Materials Park, OH: ASM International.

Warren-Findley, J. 1971. "Restoration of Equestrian Group Statues at the Memorial Bridge Plaza." Technical Report. Washington, DC: National Bureau of Standards Reports.

Watt, A., and A. Phillip. 2005. *Electroplating and Electrorefining of Metals*. Palm Springs, CA: Wexford College Press.

Wetzel, D. 1994. "Batch Hot Dip Galvanized Coatings." In *Surface Engineering,* 360–71. Vol. 5 of *ASM Handbook*. Materials Park, OH: ASM International.

Wharton, G. 1984. "Technical Examination of Renaissance Medals: The Use of Laue Back-Reflection X-ray Diffraction to Identify Electroformed Reproductions." *Journal of the American Institute for Conservation* 23:88–100.

CHAPTER 11 | CONVERSION COATINGS

The second type of inorganic coatings addressed in the present volume are conversion coatings. For metallic coatings (see chap. 10), an additional metallic layer is added on the surface. In contrast, conversion coatings modify the chemistry of the surface, converting it into a nonmetallic layer. Conversion processes include anodizing, chromating, and phosphating.

Conversion coatings are usually obtained by a controlled corrosion process in order to form, in specific conditions, a firmly bonded layer on the metal surface. They are used to promote paint bonding, to provide a decorative surface or a barrier against corrosion, and to serve as an absorptive carrier for lubricants. The conversion process can be anodic, when the part is made the anode in an electrolytic cell, or the process can be chemical, when the coating is obtained by dipping (with no potential applied). Spraying, brushing, or swabbing may also be done without the use of external current. Typical conversion coatings are oxides, chromates, phosphates, or a combination of these.

ANODIZING

Anodizing is an electrolytic oxidation process in which the surface of a metal is converted to a coherent oxide film with specific properties. Anodic coatings are employed in many functions, such as decoration, to achieve a specific texture, color, and appearance, or they may be used to prevent corrosion and wear. Anodic coatings may also be used as bases for organic or electroplated coatings or solid lubricants; they can also serve engineering functions such as abrasion-resistant surfaces, electrical insulation, and radiant heat absorption. Anodic coatings are widely applied to aluminum because its oxide is chemically stable as well as able to accept many types of useful anodic finishes. Several other metals can be anodized, including titanium and magnesium (AAC 2017).

Anodizing has been practiced since the 1920s, introduced by a chromic acid process developed for the protection of Duralumin seaplanes. The most widely used processes are those based on sulfuric acid; first patented in 1927 in the United Kingdom, this process was widely developed industrially in the United States. Other early developments include the use of oxalic acid in Japan and Germany. More recently, phosphoric acid anodizing has found significant commercial importance, particularly as a substitute for chromic acid anodizing in the aerospace industry and as a pretreatment for organic coatings. A very large number of electrolytes appear in the patent literature, but most of those used in practice are based on these four acids (Sheasby and Pinner 2001a).

Process

A typical aluminum anodizing process includes three basic steps (fig. 11.1): (a) surface preparation, (b) anodizing, and (c) sealing. The first step consists of a sequence of procedures: cleaning, etching (removing surface oxides), and desmutting (removing insoluble compounds related to aluminum alloying elements) to chemically prepare the surface in such a way that subsequent anodization proceeds uniformly over the entire surface.

The anodizing step is performed by connecting the part to the positive pole (anode) of a power supply and immersing it in the electrolyte. A current then flows, thanks to an auxiliary insoluble electrode—usually made of lead, stainless steel, aluminum, or graphite—connected to the negative pole of the power supply. During the process, oxygen (O_2) gas is released at the surface of the aluminum, and the anodized layer builds up from the bottom of the oxide film. At the same time, the top of the oxide layer dissolves chemically because of the action of the electrolyte, with which it is in contact.

In electrolytes with little or no solvent action on the oxide, coatings form rapidly. Such coatings, called barrier anodic coatings, are relatively thin and essentially nonporous, often suitable for engineering applications (fig. 11.2a). When electrolytes have appreciable solvent action on the oxide, in addition to the thin, nonporous barrier, coatings present a porous outer portion whose thickness may reach a few tens of micrometers. This structure allows a relatively high steady current flow and continued growth in coating thickness (fig. 11.2b). The process is called hard anodizing, and such coatings are termed porous-type coatings. They are employed for decorative and protective purposes, including applications requiring wear resistance.

When the anodic coating is porous, it must be sealed by immersing the anodized part in boiling water (fig. 11.1c). Hydration of the amorphous oxide is accompanied by a volume increase that effectively closes the pores. Special metal-salt-type solutions, such as nickel and cobalt acetates, and potassium dichromate solution are used as sealing solutions for some applications. To a large extent, sealing determines the performance of anodic coatings in different service environments.

Before sealing, the porous coatings may adsorb dyes or pigments to produce a full range of "impregnated" decorative colors. By this coloration method, aluminum may look like copper, pewter, or brushed bronze, and it can also be colored with brilliant blues and reds and many varieties of metallic gold and silver. Dyeing an anodic coating in several colors for decorative effects is possible using organic colorants or mineral pigments or even by electrolytic deposition in the pores (fig. 11.3). Moreover, images may be printed into the unsealed porous oxide layer using color dyes via silkscreen,

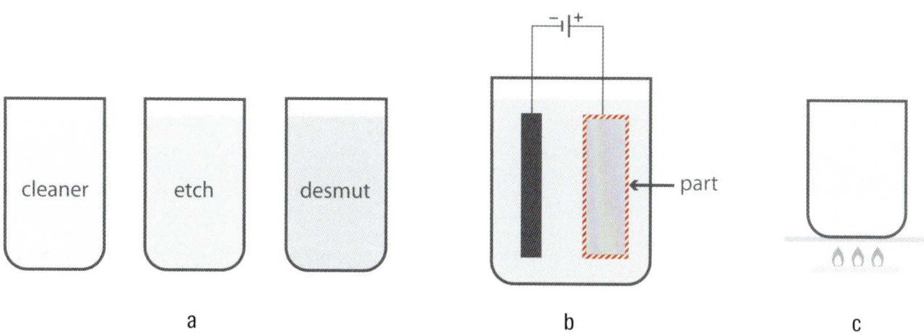

Figure 11.1 | Typical operations sequence for anodizing process: Surface preparation (a); anodizing (b); sealing (c).

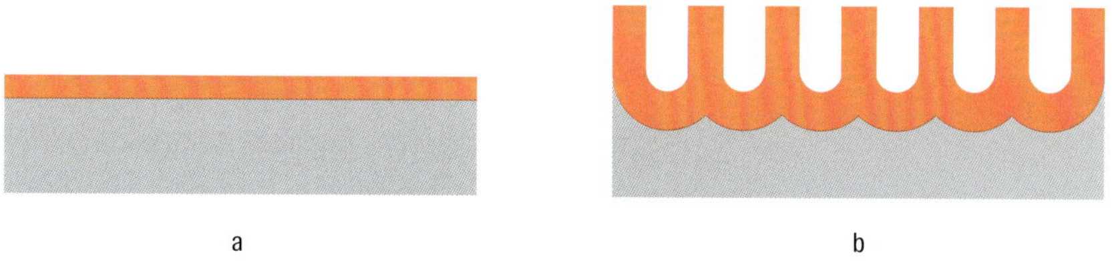

Figure 11.2 | Aluminum anodic coatings: Barrier-type (a); and porous-type (b).

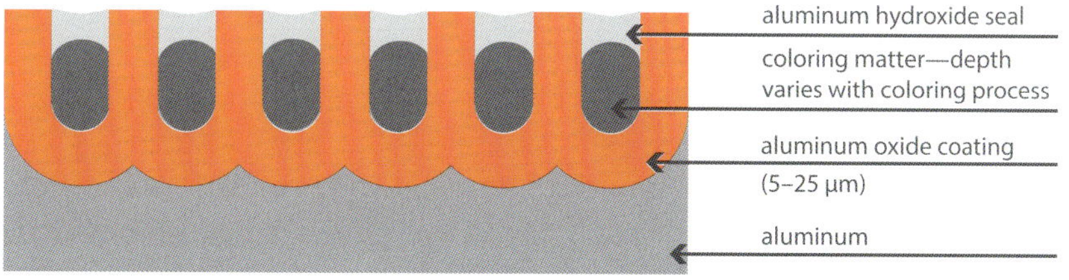

Figure 11.3 | Porous-type aluminum anodic coating with dye or color deposits, sealed.

sublimation transfer, or digital printer. Printed anodizing is sealed to prevent or reduce dye bleed-out. Uses include baseball bats, signs, furniture, surgical trays, motorcycle components, and architectural molding.

Titanium can be anodized in different colors without using dyes. The final color is a function of the oxide thickness, which is determined by the anodizing voltage. Coloring is generated by interference between the light reflected on the oxide surface and the light reflected by the underlying metal surface (fig. 11.4).

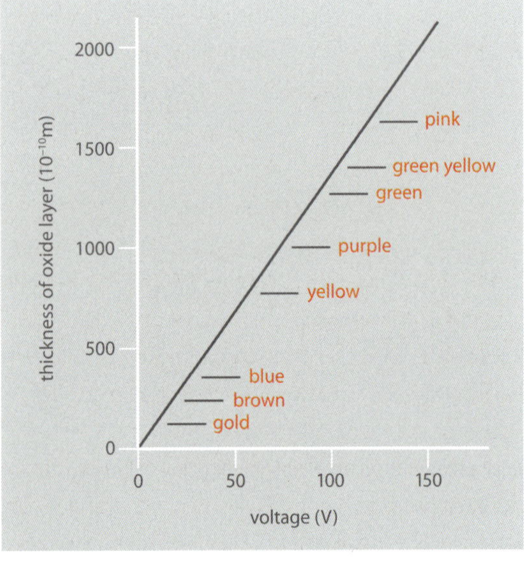

Figure 11.4 | The interference colors formed by the anodization of titanium (1.0 M sulfuric acid) related to the oxide thickness and the applied voltage (adapted from Sul, Johansson, Jeong, and Albrektsson 2001).

Properties and Applications

During the anodizing process, aluminum is progressively converted into its oxide (alumina) to form the anodized finish. Hence, the purer the material, the more homogeneous and dense will be the coating. In the case of aluminum alloys, especially those with high concentrations of alloying elements, some constituents may respond differently to anodization, being dissolved or oxidized at a rate that can be different from that of the matrix. This will affect the surface finish and color of the oxide layer and may reduce the effectiveness of the sealing process, causing reduced corrosion and wear resistance and decreasing fade resistance in dyed parts (Sheasby and Pinner 2001b). With regard to fabrication techniques, wrought alloys are generally superior to cast alloys for anodizing due to their greater homogeneity, smaller particle size of impurities, and lower porosity.

Clear anodizing is the process resulting in a translucent film, usually up to 25 μm, which provides good corrosion resistance and moderate wear resistance. The hard anodizing process creates anodic coatings with film hardness or abrasion resistance as their primary characteristic. They are usually thick (>25 μm) and are often left unsealed but may be impregnated with materials such as waxes or silicone fluids to give particular surface properties.

The type of process also plays an important role in the properties of the final product. A general view of some traditional processes and their characteristics and applications is presented in table 11.1.

Table 11.1. | Some aluminum anodizing processes in use for decorative and protective applications.

Type of process	Name or description	Aspect/color	Layer thickness (μm)	Applications
Chromic acid	Bengough-Stuart, accelerated chromic acid	Opaque gray	2–15	Mostly protection, rarely for decoration; for alloys containing <5% heavy metals
Sulfuric acid	Alumilite, Eloxal GS, hard anodizing, etc.	Transparent colorless/grayish	3–250	Hard protective coatings; good base for dyes; sealed for maximum corrosion resistance, wear resistance, and adsorptive capacity
Oxalic acid	Eloxal GX, WX, etc.	Colorless to yellow	10–65	Protection and decoration
Phosphoric acid and boric acid	Boeing process, condenser processes	Colorless, iridescent	0.2–7.5	Thin films for adhesive bonding purposes and dielectric films for electrolytic condensers

Source: Data from Sheasby and Pinner 2001a, 428.

Anodic coatings are designated by the Aluminum Association by the letter *A* followed by a two-digit number indicating the specific anodic finish. Table 11.2 gives applications of anodic aluminum coatings.

Table 11.2. | Applications of anodic aluminum coatings.

Designation	Applications
A1	General (preparation, hard coatings)
A2	Protective and decorative (<10 μm, or 0.4 mil)
A3	Architectural Class II (10–18 μm, or 0.4–0.7 mil)
A4	Architectural Class I (18 μm, or 0.7 mil)

Source: Data from AA 2003.

Characterization

When the coating is transparent, there is no simple method to detect whether an aluminum part has been anodized. A quick check can be done with a multimeter (an instrument that measures several values, usually voltage, current, and resistance) to test for the presence of any insulating barrier (the anodic coating) on the part. More specific tests (thickness, porosity, etc.) are recommended elsewhere (Sheasby and Pinner 2001b).

Regarding the conservation condition, visual examination evidences the most common defects: pitting, streaking (often related to heterogeneous metallurgical structure or physical damage that results in a variation in the anodizing response of the material), general nonuniformity of appearance, and deterioration of the appearance. Defects may originate in the substrate metal, during the various stages of production of the coating, or in the service environment. A visual guide is presented elsewhere (Short 2003).

In the case of outdoor parts, pitting generally occurs underneath grime deposits on faces protected from the natural washing of rain. The anodic film is destroyed locally, resulting in the accumulation of associated corrosion products and a circular stain around individual pits.

Regarding maintenance, parts exposed outdoors must be included in a program that ensures regular removal of dirt, grime, and the like. Any chemical attack on a decorative surface that is not regularly cleaned becomes increasingly severe, and after several years, the finish may prove impossible to restore. Areas protected by an overhang that are not washed by rain will tend to deteriorate faster than surfaces that are regularly washed (see fig. 4.6). Regular washing with water plus a suitable wetting agent will maintain the appearance of the finish and prevent pitting. Abrasive, acidic, or alkaline cleaners may destroy the finish.

CHROMATE CONVERSION COATINGS

Chromate conversion coatings, often referred to as chromate coatings, are thin amorphous inorganic films formed on metal surfaces by reaction with aqueous solutions whose main film-forming agent is a soluble chromate species. Such coatings form spontaneously without application of current or voltage and consist of mixed metal oxides and hydroxides. They are most commonly applied to aluminum, zinc, and cadmium and their alloys (Buchheit and Hughes 2003). These coatings are primarily used to improve the adhesion of subsequent organic coatings or to impart corrosion resistance during atmospheric exposure.

The earliest conversion coatings were developed for aluminum alloys, which were just emerging as engineering materials in the early 1900s. Cu-bearing aluminum alloys, which presented an excellent combination of low density and high strength but low corrosion resistance, needed additional protection. To this end, chromate coatings were applied as a base for paint and provided good protection. During World War II, demand for the protection of metal parts—in particular, zinc- and cadmium-plated ones in tropical service—stimulated widespread use of chromate finishing, because it prevents the development of a bulky white corrosion product called "white rust" (a characteristic oxidation product of Zn and Cd), which is particularly deleterious when electrical contact is needed (Buchheit 2002; Burns and Bradley 1967). After the war, chromate finishing was adapted to many commercial manufacturing processes.

Process

The application of chromate coatings is performed by immersion in chromating solutions at 15°C–32°C for periods ranging from a few seconds to several minutes. The bath compositions are largely proprietary, but all contain two basic constituents: hexavalent chromium ions and a mineral acid. Some also contain one or more organic acids (Burns and Bradley 1967).

Upon immersion, the part is attacked by the mineral acid, which leads to a local rise in pH at the interface. Above a critical pH, a thin, complex chromium-metal gel precipitates on the surface. Chromate films are generally noncrystalline, nonporous, and gel-like after formation. As the coating dries or ages, it slowly hardens and becomes hydrophobic, less soluble, and more abrasion resistant. It should age for twelve to twenty-four hours after formation before being subjected to normal handling (Innes 1984).

Properties and Applications

There is general agreement that chromate coatings consist primarily of a mixture of Cr(III) and Cr(VI) compounds and also include other bath ingredients and elements from the substrate. The precise composition of chromate coatings depends, therefore, on the alloy substrate, bath formulation, processing parameters, and coating age.

The amount of coating deposited is generally expressed in grams per unit of surface area, which is referred to as *coating weight*. It increases with immersion time and solution temperature. Typical values range from 0.16 to 1.6 g/m^2 (15–150 g/ft^2).

Chromate coatings are available in a wide range of colors such as olive green, bronze, iridescent yellow, blue bright, and clear (table 11.3; fig. 11.5). The final color depends on the type of chromating solution used and the thickness of the coating. Thinner coatings are usually lighter in color.

The degree of protection is improved by increasing both the thickness and the chromate content in the film. Generally, thin, clear, bright coatings impart the least corrosion protection; thick olive green and bronze confer the most; and iridescent yellow films fall in the middle. Because the chromate content is the attribute that determines the ability to store and release Cr(VI) into an attacking solution, it is key to the self-healing property, which may be defined as the ability of a chromate conversion coating to heal small chemical or mechanical defects that could serve as initiation sites for corrosion (Buchheit and Hughes 2003). When used as a base for painting, chromate conversion coatings not only provide a good bonding surface but also interpose a nonporous protective barrier between the paint and the metal.

Once dried, coatings withstand considerable handling, stamping, and cold forming, but they cannot resist sharp scratching or harsh or continuous abrasion. The electrical resistance is low compared with other types of nonmetallic finishes, a property that enables the use of chromate coatings to protect electrical contacts and connectors without increasing surface resistance.

Chromate films are often applied on Zn and Cd plated steel in order to improve the decorative appearance by preventing white rust and increasing resistance to finger staining during handling. The treated work retains its brilliance longer in mild exposure conditions, significantly increasing shelf life. Chromate conversion films are also applied on aluminum parts that will subsequently be coated, in order to improve paint adherence (fig. 11.6).

Chromate conversion coatings find extensive use in low-cost items where high abrasion resistance is not necessary. Bright finishes have wide application on objects such as tubular furniture, tools, nuts and bolts, radio and television chassis and loudspeakers, small components of home appliances, and cars. Dyed finishes are produced in a supplemen-

Table 11.3. | Characteristics of chromate coatings.

Color	Thickness (µm)	Cr(VI) mg/m^2
Blue	0.025–0.08	10–30
Yellow	0.25–0.50	60–220
Olive green	1.0–1.5	300–360

Source: Data from Carla Martini, associate professor, University of Bologna.

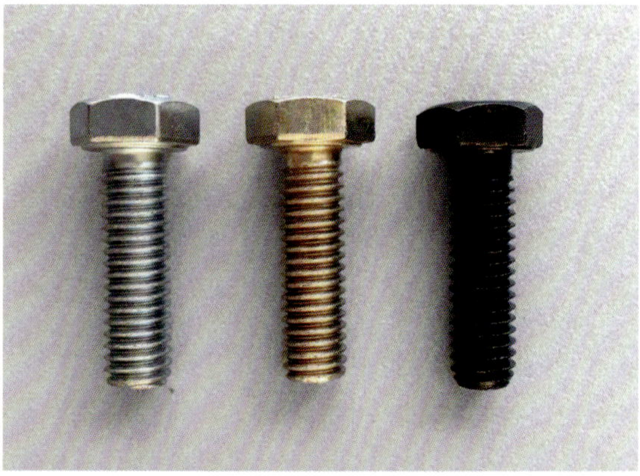

Figure 11.5 | Zinc-plated bolts with different color chromates: clear, yellow, and black. Photo: Virginia Costa.

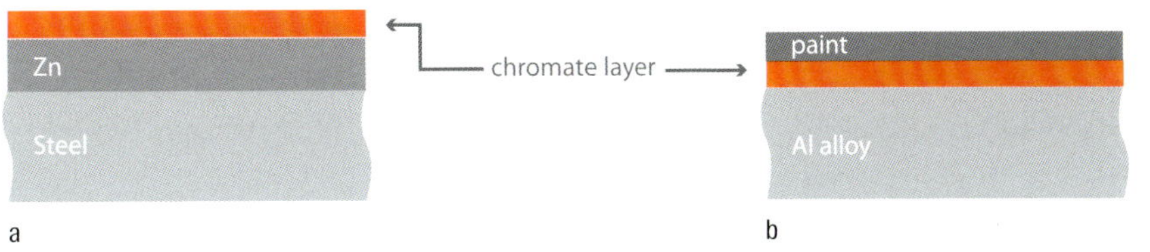

Figure 11.6 | Chromate used as a top coating to improve appearance and corrosion resistance (a), and as an intermediate layer to improve the adhesion of the top paint layer (b) (reproduced from Carla Martini).

tal processing step that follows the chromate coating. They are used for color coding (e.g., identification of fasteners), for blackening camera parts, and for simulating brass in decorative applications such as lamps and furniture.

Characterization

A pale yellow or blue surface color is a good indication of the presence of a chromate coating. For the most part, if no color can be observed, there is no chromate film. However, this does not necessarily mean that chromates had not been applied, because they generally last only up to six weeks on galvanized coatings, and they degrade even more quickly in very wet or humid environments.

There are a number of specifications for testing for the presence of chromate coatings and identifying the type (ASTM International 2014). These are usually spot tests, and they are not simple to implement because the solutions employed must be mixed just prior to using, and the surface has to be free of secondary supplementary coatings, such as oil-, water-, and solvent-based polymers or wax (Bahadori 2015, 220–23).

PHOSPHATE CONVERSION COATINGS

Phosphate conversion coatings are conversion coatings for which a dilute solution of phosphoric acid and phosphate salts is applied by immersion or spray. A chemical reaction with the surface of the part leads to the formation of a layer made of insoluble crystalline phosphates. The treatment can be applied to iron, steel, galvanized steel, aluminum, and electrodeposited zinc and cadmium. Thanks to their particular composition and structure, phosphate coatings are used as a pretreatment prior to applying organic coatings, as temporary protection against corrosion, and as an efficient lubricant carrier or antiwear coatings.

Phosphate coatings have been used for protecting steel surfaces since the beginning of the twentieth century. The first important commercial application of phosphate coatings followed the proposal of T. W. Coslett in 1908 to treat iron and steel parts with a solution of phosphoric acid and iron fillings (Burns and Bradley 1967). Other formulations based upon zinc (1909) and manganese (1911) phosphate were also developed, and some were given different names, such as Parkerizing (1914) and Bonderizing (1929). Bonderizing employed the addition of accelerators (Cu, Ni), which drastically reduced the processing time. Since that time, numerous developments have been introduced (Sankara Narayan 2005).

Process

As with many surface treatments, the phosphating sequence depends upon the surface condition of the base metal and may comprise several steps, including pretreatment (degreasing, pickling, rinsing), phosphating, and posttreatment (rinsing, sealing, drying).

A wide range of phosphating compositions is available. In essence, they are all dilute solutions based on phosphoric acid that contain alkali metal or heavy metal ions in addition to suitable accelerators (Sankara Narayan 2005).

All phosphate coatings are produced by the same type of chemical reaction: the acid bath reacts with the surface of the metal to be coated, neutralizing the solution at the solid-liquid interface. As a consequence, crystals of metal phosphates precipitate on the surface from the solution. In fact, crystals are attracted to the surface of the metal by the normal electrostatic potential and are deposited on the cathodic sites.

Phosphate deposition can be achieved through the use of both immersion and spray processes. The choice of method depends upon the size and shape of the substrate to be coated. The immersion process produces uniform coatings. The spray process is preferred when shorter processing times are required. Occasionally, a surface may be coated by brushing or wiping.

Properties and Applications

Phosphate coatings consist of layers of water-insoluble metal phosphates of varied crystal size tightly bonded to the surface (fig. 11.7). The chemical composition and crystal structure of the coating depends on the metal to be coated, on its surface finishing, and on the type of bath used. When fully developed, the coating will cover the surface completely (Canning & Co. 1982).

Phosphate coatings are available in three types: iron, zinc, and manganese phosphate (table 11.4). Each of these coatings has its own unique properties—the physical structure of the coating being an important consideration for particular applications (Douty, Stockbower, and Jones 1984). Phosphate coatings can range from less than 0.1 to over 2 mils in thickness. Industry standards usually express the coating requirements by weight per square meter or milligrams per square foot, rather than by the thickness of the deposit. Coating weight requirements are generally dependent upon the type of phosphate process involved.

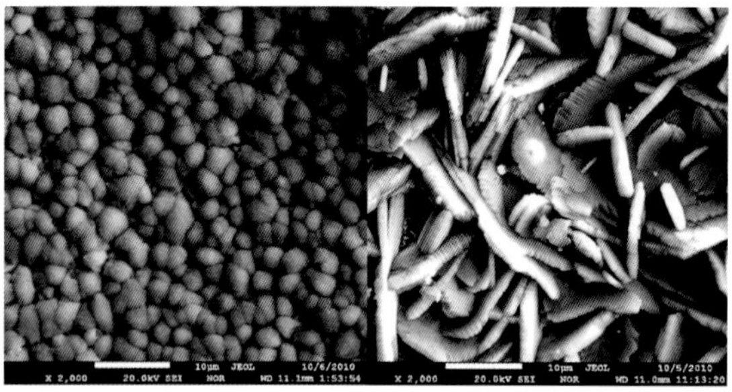

Figure 11.7. | SEM micrograph of the zinc phosphate crystals formed on steel surface: globular-shaped phosphate crystals of 2–6 µm (a); plate-like structure with larger crystal size of 25–40 µm (b). The basic constituents of the solution were the same in both cases, but differ in bath parameters (acidity and accelerator concentration) (reproduced from Rani et al. 2012, fig. 1).

Bare phosphate coatings are seldom employed. For the most part, they serve to promote the adhesion of organic coatings to metal substrates and thus enhance corrosion resistance. Phosphate coatings with supplementary lubrication are used to assist in cold deformation processes such as wire drawing, cold heading, and impact forming (Douty, Stockbower, and Jones 1984).

Characterization

Phosphate-coated parts usually have a dull grayish or blackish, fairly nonmetallic appearance. Zn and Mn phosphate coatings range from gray to black and present a uniform crystalline texture, generally recognizable under optical microscopy at low magnification. Iron phosphate coatings have no evident crystalline texture; rather, they appear to be amorphous, with color varying from blue to brown. Tests for the presence of a phosphate coating include spot tests (Bahadori 2015, 213–19). Shiny areas of the coating indicate a locally defective condition where the phosphate coating has not formed because of surface contamination.

CONSERVATION-RELATED LITERATURE

Publications on conversion coatings in the conservation field are rare. Three have been found that deal with historic finishes for architectural aluminum (Flandro and Thomas-Haney 2015), the conservation of aluminum theater masks (Camarero Heras and Ruiz Sánchez 2011), and the use of rust converters to protect outdoor iron artifacts (Binnie et al. 1995).

Table 11.4. | Properties and applications of phosphate coatings.

Phosphate coating	Thickness mg/m^2 (mg/ft^2)	Properties	Applications
Fe	0.3–1.0 (30–100)	Thin amorphous coating; needs posttreatment; improved adhesion of paints to the substrate	Base coating for paint or for bonding other materials
Zn	1–30 (100–3000)	Crystalline structure; retains rust-preventive oils and waxes; uniform coating; good corrosion protection by bonding the final finish to the treated material	Springs, fasteners, pulleys
Mn	10–30 (1000–3000)	Large crystalline structure; excellent wear resistance; excellent oil-retentive properties	Bearings, piston rings, gears, cylinder liners, internal combustion parts

Source: Data from Canning W. and Co.1982, 842–44.

REFERENCES AND BIBLIOGRAPHY

AA [Aluminum Association]. 2003. *Designation System for Aluminum Finish*. DAF-45. Washington, DC: Aluminum Association.

AAC [Aluminum Anodizers Council]. 2017. "What is Anodizing? Anodizing . . . The Finish of Choice." Wauconda, IL: Aluminum Anodizers Council. *https://www.anodizing.org/page/what_is_anodizing?*

ASTM International. 2014. *ASTM B201. Standard Practice for Testing Chromate Coatings on Zinc and Cadmium Surfaces*. West Conshohocken, PA: ASTM International.

Bahadori, A. 2015. *Essentials of Coating, Painting, and Lining for the Oil, Gas and Petrochemical Industries*. Amsterdam: Gulf Professional Publishing / Elsevier.

Binnie, N. E., L. S. Selwyn, C. Schlichting, and A. Rennie-Bisaillion. 1995. "Corrosion Protection of Outdoor Iron Artifacts Using Commercial Rust Converters." *Journal of the IIC-CG* 20:26–40.

Buchheit, R. G. 2002. "Conversion Coating Science and Technology: Is It Evolving or Is It Stuck?" In *Corrosion Science: A Retrospective and Current Status in Honor of Robert P. Frankenthal*, edited by G. S. Frankel, H. S. Isaacs, J. R. Scully, and J. D. Sinclair, proceedings vol. 2002-13, 430–40. Pennington, NJ: Electrochemical Society.

Buchheit, R. G., and A. E. Hughes. 2003. "Chromate and Chromate-Free Conversion Coatings." In *Corrosion: Fundamentals, Testing, and Protection*, 720–35. Vol. 13A of *ASM Handbook*. Materials Park, OH: ASM International.

Burns, R. M., and W. W. Bradley. 1967. *Protective Coatings for Metals*. New York: Reinhold.

Camarero Heras, C., and A. Ruiz Sánchez. 2011. "Restauración de las máscaras de alumínio procedentes del Teatro Fleta de Zaragoza. Parte I: Intervención en las máscaras situadas en el escenario del teatro." *Kausis* 8:21–29.

Canning W. and Co. 1982. "Phosphating Processes." In *The Canning Handbook: Surface Finishing Technology, Integrated Design*. 23rd ed., 840–84. Birmingham: Kluwer Academic Publishers.

Delta Specialty Coatings. 2017. "Capabilities." Houston: Delta Specialty Coatings. *http://www.deltaspecialtycoatings.com/capabilities.php*.

Douty, A., E. A. Stockbower, and W. C. Jones. 1984. "Phosphate Coating Processes." In *Electroplating Engineering Handbook*, edited by L. J. Durney, 366–73. New York: Van Nostrand Reinhold.

Flandro, X., and H. M. Thomas-Haney. 2015. "A Survey of Historic Finishes for Architectural Aluminum, 1920–1960." *APT Bulletin* 46 (1): 13–21.

Innes, W. P. 1984. "Chromate Conversion Coatings." In *Electroplating Engineering Handbook*, edited by L. J. Durney, 390–95. New York: Van Nostrand Reinhold.

Rani, N., et al. 2012. "Optimization of Phosphate Coating Properties on Steel Sheet for Superior Paint Performance." *Journal of Coatings Technology and Research* 9 (5): 629–36.

Runge, J., and L. Chesterfield. 2010. "The Science of Successfully Anodizing Die Cast Substrates." *Aluminium International Today* [*AIT*], Nov.–Dec.: 30–36.

Sankara Narayan, T. S. N. 2005. "Surface Pretreatment by Phosphate Conversion Coatings: A Review." *Reviews on Advanced Materials Science* 9:130–77.

Sheasby, P. G., and R. Pinner. 2001a. "Decorative and Protective Anodizing." In *The Surface Treatment and Finishing of Aluminum and Its Alloys*, 6th ed., edited by P. G. Sheasby and R. Pinner, vol. 2, 427–46. Materials Park, OH: ASM International.

———. 2001b. "Properties and Tests of Anodic Oxide Coatings." In *The Surface Treatment and Finishing of Aluminum and Its Alloys*, 6th ed., edited by P. G. Sheasby and R. Pinner, 963–1111. Materials Park, OH: ASM International.

Short, T. 2003. *The Identification and Prevention of Defects on Anodized Aluminium Parts*. Stevenage, UK: Metal Finishing Information Services. *https://www.fot.de/uploads/docs/fehlermoeglichkeiten.pdf*.

Sul, Y.-T., C. B. Johansson, Y. Jeong, and T. Albrektsson. 2001. "The Electrochemical Oxide Growth Behavior on Titanium in Acid and Alkaline Electrolytes." *Medical Engineering and Physics* 23:329–46.

GLOSSARY

accelerated corrosion test. A method designed to approximate, in a short time, the deteriorating effect on a metal under normal long-term service conditions.

active metal. A metal ready to corrode or undergoing corrosion.

addition agent. A substance added to a solution for the purpose of altering or controlling a process.

age hardening. The hardening of a metallic alloy by aging, usually after rapid cooling or cold working.

aging. A change in the properties of certain metals and alloys that occurs at ambient or moderately elevated temperatures after heat treatment.

alloy. A substance with metallic properties that is composed of two or more chemical elements, at least one of which is an elemental metal.

alloy plating. The codeposition of two or more metallic elements.

alloy system. A complete series of composition produced by mixing in all proportions any group of two or more components, at least one of which is a metal.

alloying element. An element added to and remaining in a metal that changes its structure and properties.

alpha iron. The solid phase of pure iron, stable below 910°C, possessing a body-centered cubic (BCC) lattice; it is ferromagnetic at room temperature.

aluminizing. The formation of an aluminum or aluminum alloy coating on a metal by hot dipping, hot spraying, or diffusion.

anion. A negatively charged ion that migrates through the electrolyte toward the anode under the influence of a potential gradient.

annealing. A treatment consisting of heating to and holding at a suitable temperature, followed by cooling at a suitable rate, used primarily to soften metallic materials but also to simultaneously produce desired changes in other properties or in microstructure.

anode. The electrode in an electrochemical cell or galvanic couple at which oxidation occurs.

anodizing. The production of a conversion coating on a metal surface by anodic oxidation, most frequently applied to aluminum.

binary alloy. Any specific alloy composition in a series of compositions produced by mixing a pair of components in all proportions.

body-centered cubic (BCC). A common crystal structure found in some elemental metals in which, within the cubic unit cell, atoms are located at corner and center positions.

brush plating. A portable process for applying localized electroplated deposits and anodized coatings to metals. Also known as selective plating.

cast structure. The metallographic structure of a casting, evidenced by the shape and orientation of the grains as well as by the segregation of impurities.

cathode. In an electrochemical cell, the component upon which reduction takes place.

cathode efficiency. The current efficiency at the cathode.

chromating. A treatment of metal in a solution of hexavalent chromium compounds to produce a conversion coating.

cold working. The plastic deformation of metal under conditions of temperature and strain rate that induce strain hardening.

conversion coating. A coating consisting of a compound of the surface metal, produced by chemical or electrochemical treatment.

creep resistance. The time-dependent permanent deformation of a material that occurs under stress.

crystal structure. For crystalline materials, the manner in which atoms or ions are arrayed in space.

dendrite. A crystal with a treelike branching pattern; most evident in cast metals slowly cooled through the solidification range.

dezincification. Corrosion in which zinc is selectively leached from zinc-containing alloys; it occurs most commonly in copper-zinc alloys.

die casting. A permanent mold casting process in which liquid metal is injected into a metal die under high pressure. This high-rate production process requires expensive equipment and precision-matched metal dies.

differential aeration cell. An electrolytic cell, the electromotive force of which is due to a difference in air (oxygen) concentration at one area as compared to that at another region of the same material.

dislocation. A linear crystalline defect around which there is atomic misalignment.

drawing. A forming technique used to fabricate metal wire and tubing by pulling the material through a die by means of a tensile force applied to the exit side.

ductility. The measure of a material's ability to undergo appreciable plastic deformation before fracture.

electrochemical reaction. A reaction consisting of a flow of electrons between cathodic and anodic areas on metallic surfaces.

electrode. An electronic conductor used to establish electrical contact with an electrolytic part of a circuit.

electroforming. The production or reproduction of an article by electrodeposition upon a mandrel or mold that is subsequently separated from the deposit.

electroless plating. A process in which metal ions in a dilute aqueous solution are plated onto a substrate by means of autocatalytic chemical reduction.

electrolyte. A solution through which an electric current may be carried by the motion of ions.

electromotive force series. A list of elements arranged according to their standard electrode potentials, with noble metals being positive and active metals being negative.

electroplating. The electrodeposition of a metal or alloy in an adherent form on an object serving as a cathode.

electrotype. A copy made of an original artwork by depositing a metal (usually copper) either directly onto the surface of the original or into a mold taken from the original.

exfoliation. Corrosion proceeding laterally from the site of initiation along planes parallel to the surface, creating a layered appearance.

extrusion. A forming technique in which a material is forced by compression through a die orifice.

face-centered cubic (FCC). A common crystal structure found in some of the elemental metals in which, within the cubic unit cell, atoms are located at all corner and face-centered positions.

filiform corrosion. Corrosion that occurs on a metal under some coatings in the form of randomly distributed threadlike filaments.

forging. The mechanical forming of a metal by heating and hammering.

galvanic corrosion. Accelerated corrosion of a metal due to an electrical contact with a more noble metal or a nonmetallic conductor in a corrosive electrolyte.

galvanic series. A list of metals and alloys arranged according to their relative corrosion potentials in a given environment.

galvanizing. The coating of a metal surface with zinc using any of various processes.

grain boundary. The interface separating two adjacent grains with different crystallographic orientations.

grain size. The measure of the area or volume of a grain in a polycrystalline metal or alloy, usually expressed as an average when the individual sizes are uniform; it is reported as the number of grains per unit area or volume.

hardness. The resistance of a material to deformation by surface indentation or abrasion.

heat treatment. The heating and cooling of a metal or alloy in a manner, with regard to temperature and time, that is designed to obtain useful combinations of properties.

hexagonal close packed (HCP). A crystal structure found in some metals. The unit cell is of hexagonal geometry and is generated by the stacking of close-packed planes of atoms.

hot-dip coating. A metallic coating obtained by dipping the base metal into a molten metal.

immunity. A state of resistance to corrosion due to the thermodynamic stability of a metal or alloy.

lost-wax process. A casting method in which a ceramic slurry is applied around a wax pattern and hardened to make a mold; the pattern is removed by heat, and the molten alloy is poured into the space formerly occupied by the pattern.

macrostructure. The general crystalline structure of a metal or alloy and the distribution of impurities as seen on polished and etched surfaces, either by the naked eye or under magnifications of less than 10 diameters.

metallic bond. An interatomic bond involving the nondirectional sharing of nonlocalized valence electrons that are mutually shared by all the atoms in the metallic solid.

microhardness. Hardness measured with an indenter under the application of a low load, usually between 1 g and 200 g.

microstructure. The structural features of an alloy (e.g., grain and phase structure) that can be observed under a microscope.

modulus of elasticity. The ratio of stress to strain when deformation is completely elastic; also, a measure of the stiffness of a material.

passivity. The loss of chemical activity by some active metals and alloys under particular environmental conditions.

phase. A physically homogeneous and distinct portion of a material system.

phase diagram. A graphic representation of the relationships among temperature, composition, and regions of phase stability, ordinarily under conditions of equilibrium.

phosphating. The formation of an adherent phosphate coating on a metal by immersion in a suitable aqueous phosphate solution.

pitting corrosion. A form of very localized corrosion of a metal surface, confined to a point or small area, that takes the form of cavities.

plastic deformation. The permanent (inelastic) distortion of a metal under applied stress that strains the material beyond its elastic limit.

poultice corrosion. Corrosion due to the collection of debris and salts on metallic parts that are kept moist by weather.

Pourbaix (potential-pH) diagram. A plot of the redox potential of a corroding system versus the pH of the system; the diagram shows regions within which the metal itself—or some of its compounds—is stable.

precipitation hardening. Hardening caused by precipitation of a constituent from a supersaturated solid solution.

quenching. The rapid cooling of metals from a suitable elevated temperature.

recrystallization. The formation of a new, strain-free grain structure from the structure existing in cold-worked metal.

reference electrode. An electrode with a known and highly reproducible potential.

rolling. A metal-forming operation that reduces the thickness of a sheet stock.

salt spray test. An accelerated corrosion test in which specimens are exposed to a fine mist of a solution usually containing sodium chloride. Also known as salt fog test.

sand casting. A casting method in which a molten metal is poured into a cavity shaped inside a body of sand and allowed to solidify.

segregation. The nonuniform distribution of impurities or alloying elements.

selective plating. See *brush plating*.

slip. Plastic deformation as the result of dislocation motion; also, the shear displacement of two adjacent planes of atoms.

solid solution. A homogeneous crystalline phase containing two or more chemical species in concentrations that may vary between limits imposed by phase equilibrium.

solubility limit. The maximum concentration of solute that may be added without forming a new phase.

solute. One component or element of a solution that is dissolved in the solvent and is present in a minor concentration.

solution heat treatment. The process used to form a solid solution by dissolving precipitate particles.

stress relief. A heat treatment for the removal of residual stresses.

temper designation. A letter-digit code that designates the mechanical and/or thermal treatment to which a metal alloy has been subjected.

tempering. Heat treatment process for steels performed after quenching.

tensile strength. The highest load applied in breaking a tensile test piece divided by the original cross-sectional area of the test piece.

throwing power. The ability of a plating solution to produce a uniform metal distribution on a cathode of irregular shape.

toughness. The ability of a metal to absorb energy and deform plastically before fracturing.

work hardening. An increase in the hardness and strength of a metal or alloy caused by plastic deformation at temperatures below the recrystallization range.

yield strength. The stress at which a material exhibits a specific deviation from proportionality of stress and strain; an offset of 0.2% is used for many metals.

ABOUT THE AUTHOR

Virginia Costa qualified as an engineer in metallurgy in Brazil and earned her PhD from the Technical University of Berlin. She spent twenty years as a professor in the area of corrosion and protection of metals at the Federal University of Rio Grande do Sul in Porto Alegre, Brazil. She then became a consulting expert at the international level and for the past twenty years has applied her scientific and technical background to the conservation of metallic cultural heritage. During this time, in addition to being a lecturer at the Institut National du Patrimoine, Paris, she has worked for several institutions: the Musée de la Musique and the Laboratoire de Recherche des Monuments Historiques, Paris; the Royal Institute for Cultural Heritage (KIK-IRPA), Brussels; Artesis Plantijn, Antwerp; the Kunsthistorisches Museum, Vienna; English Heritage, London; and the Museu Paulista, São Paulo. She was awarded an ICCROM fellowship in 2009 and a Getty Museum scholarship in 2012. Costa has written articles, given numerous invited lectures, and organized several training courses on the uses of electrochemistry and metallography for the analysis and conservation of metals.

INDEX